SpringerBriefs present concise summaries of cutting-edge research and practical applications across a wide spectrum of fields. Featuring compact volumes of 50 to 125 pages, the series covers a range of content from professional to academic. Typical topics might include:

- A timely report of state-of-the art analytical techniques
- A bridge between new research results, as published in journal articles, and a contextual literature review
- A snapshot of a hot or emerging topic
- An in-depth case study
- A presentation of core concepts that students must understand in order to make independent contributions

More information about this series at http://www.springer.com/series/10207

Malte C. Ebach • Bernard Michaux

Biotectonics

Tectonics as the Driver of Bioregionalisation

 Springer

Malte C. Ebach
University of New South Wales
Kensington, NSW, Australia

Bernard Michaux
Kaukapakapa, New Zealand

ISSN 2192-8134 ISSN 2192-8142 (electronic)
SpringerBriefs in Evolutionary Biology
ISBN 978-3-030-51772-4 ISBN 978-3-030-51773-1 (eBook)
https://doi.org/10.1007/978-3-030-51773-1

This Springer imprint is published by the registered company Springer Nature Switzerland AG
The registered company address is: Gewerbestrasse 11, 6330 Cham, Switzerland

Prologue

What This Book Is About

Tectonic plates are constantly moving, either crashing into one another creating a mosaic of mountains and shallow seas, tearing apart and isolating large swathes of land, or sliding past each other translating areas laterally. In each case, plate tectonics separates populations leading to the evolution of new species. Tectonics is also responsible for the destruction of life, for instance when large coral reefs or shallow seas are compressed during periods of mountain building. Could recent research into these processes provide enough evidence to show that tectonics may be the ultimate driver of life on Earth?

Our book will delve into the current research in tectonics, particularly neotectonics, and its impact on rapid changes in biogeographical classification, also known as bioregionalisation. We also introduce a new term *biotectonics*, which studies the impact of tectonics on biogeoregionalisation. The question we will ask is how tectonics directly influences the distribution of biota in three case studies: the Palaeozoic, Mesozoic and early Palaeogene Australides orogeny, which spans the Proto-Pacific coast of South America, Antarctica and Australasia; the Neogene of Australia; and the Neogene of the Amazon Basin. To conclude, we will discuss the role of tectonic extinction on bioregionalisation.

Is Tectonics the Ultimate Driver of Life on Earth?

Recently, there has been much discussion about plate tectonics driving the nutrient cycle since the Neoproterozoic (Large et al. 2015), although there has been little discussion about its impact on biodiversity and biological distributions. The physical effects of shaping a landscape makes tectonics a significant factor in driving biotic distributions, which in turn create new natural areas (an area or group of areas that share a common geological and biological history (Michaux 2010)) and form

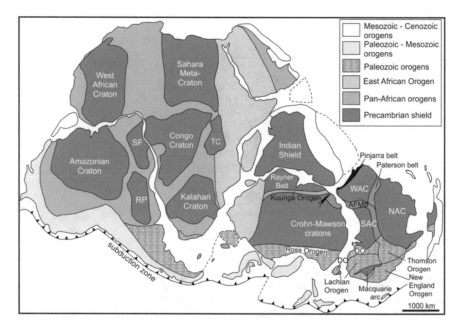

Fig. 1 Australides map. Gondwana at the close of the Palaeozoic following the reconstruction (from Fergusson and Henderson 2015, Fig. 1) Abbreviations: *AFMB* Albany-Fraser-Musgrave belt, *DO* Delamerian Orogen, *NAC* North Australian Craton, *RP* Río de la Plata Craton, *SAC* South Australian Craton, *SF* São Francisco Craton, *TC* Tanzania Craton, and *WAC* West Australian Craton

monophyletic groups on an areagram (Ebach and Michaux 2017). The slow collisions between tectonic plates have rapid knock-on effects further afield. A subducting plate may, for example, change the direction of a nearby mantle plume creating a down-welling, which creates a significant change in the topography of the overriding plate many hundreds or even thousands of kilometres from the active collision zone. The changes in plate topography and the formation of tectonostratigraphic terranes impact continental and marine biotic distributions (Chap. 1). We call this process of Earth dynamics and biotic evolution *Biotectonics*.

The large-scale effects of biotectonics can be seen in the changes in geography and biogeographical distributions during the Australides orogeny (Fig. 1). Present-day distributions of Gondwanan taxa reveal an older pan-continental distribution of fauna and flora that was broken up by large-scale terrane movement across parts of the Australides (Vaughan et al. 2005). The further back in time one looks, the larger terrestrial biogeographic regions become along the Proto-Pacific margin (Chap. 2). The difference between modern day and Mesozoic distributions is the role that cratons play in crustal evolution. Cratons form stable plateaus with a contrasting geomorphology with respect to the highly dynamic basin/arc morphologies found on the outboard terranes. These two very different parts of the plate are host to very different biogeographic areas.

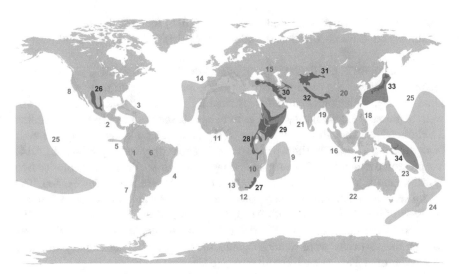

Fig. 2 Biodiversity hotspots (Myers et al. 2000) and the underlying tectonic activity. 1 and 7, South American and Nazca plate collision; 2, Cocos and Caribbean plate collision; 3, Caribbean, North, and South American plate collision; 5, Nazca and South American plate collision and hotspot under Galapagos Is.; 8, Juan de Fuca and North American plate collision; 14, African and Eurasian plate collision; 16–17, Australian and Sunda plate collision; 15, 20, 30–32, Persia-Tibet-Burma Orogen (Lui and Bird 2008); 18, Philippine and Sunda plate collision; 19, Transform boundary between Sunda and Burma plate, Eurasian and Sunda plate collision; 23, 24, 34, Australian and Pacific plate collision; 28 and 29, East African Rift; 33, Okhotsk, Philippine, and Pacific plate collision. Only 8 areas (i.e. 4, 6, 9–13, 22 and 27) lie on passive tectonic margins, however, in the Western Ghats (21) lies the active Thenmala fault system, and several have speculated that Sri Lanka (21) may be tectonically active. Southwestern Australia (22) is seismically and neotectonically active and has undergone uplift due to the continental tilt; the large Polynesia and Micronesia (25) area covers numerous tectonically active zones. Creative Commons CC BY-SA 3.0, Wikipedia

Plate collisions may also have effects farther afield. The Australian plate has been thrust under the Pacific and Sunda plates creating what Sandiford (2007) called a "Tilting Continent" (see Chap. 3) in which the northern half of Australia lies 200 meters lower than it should. The tilt resulted in marine transgressions across the northern margin, which was composed of a broad and shallow shelf, and provided the drainage and soils necessary for the growth of lush vegetation in stark comparison to the uplifted Nullarbor Plain at the southern edge of the continent. The constant tilting between the north and south resulted in the drowning of the Nullarbor Plain and Murray basin in the Miocene (20 Ma) biologically isolating east Australia from the west Australia. Also the dynamic topography resultant of a down-welling under Lake Eyre has created a lower and wetter eastern desert basin and a higher and direr western desert plateau. This east-west pattern can still be seen today in the Eucalypts and *Acacia* (Murphy et al. 2019; Ebach and Murphy 2020). In fact, tectonic activity leads to the dynamic topography that seems to coincide with current biodiversity hotspots (Fig. 2), which Müller and colleagues (2016) found both along the eastern coastline and Barrent-Morre and colleagues (2015) found in the southwest of Australia.

Plate tectonics is not the only geographical driver: the convecting plumes in the mantle that power the plate tectonic conveyor belt also create undulations in the plate surface, a phenomenon that is referred to as dynamic topography. A downwelling plume pulls the flimsy plate to form basins, such as Lake Eyre in Australia. Sandiford (2007) has shown that down-welling may alter a landscape by hundreds of meters in geologically short periods. For example, the work by Flament (2014) has presented evidence that the Amazon flowed towards the Pacific and emptied into a vast inland sea about 14 million years ago, well before the Andes existed. The inland sea was created by a strong down-welling, which abruptly stopped with the collision of the Pacific and South American plates and the rise of the Andes. The Amazon's subsequent change in flow direction meant that many of the tributaries feeding the vast river were running against the main current, effectively isolating freshwater organisms at each confluence (see Chap. 4).

Plate tectonics may well be a driver of evolution, but it also drives extinction. During the Late Permian and Early Triassic (~250 Ma), which saw a loss of approximately 90% of all species, continental landmasses converged to form the supercontinent Pangaea. Wignall (2017) noted the correlation between the formation of Pangaea and the largest ever mass extinction, as well as the eruption of large-scale volcanism. Capriolo et al. (2020) linked the end-Triassic extinction event (~200 Ma), which marks the boundary between the Triassic and Jurassic, to the rapid and extensive outgassing of CO_2 that accompanied the eruption of the Central Atlantic Magmatic Province. These authors suggested that the extreme global warming and ocean acidification that resulted from this rapid increase in atmospheric CO_2 was responsible for this extinction event. Areas of high diversity, such as coral reefs, shallow seas, inland seas and estuaries, all occur along coastlines, which are exactly the environments that undergo severe reduction during the formation of a supercontinent as separate continents amalgamate.

The collision of North with South America, Africa with Europe, India with Eurasia, and Australia and the Pacific with Eurasia had all started by 40 Ma. Greater India, which was once connected to Madagascar and Australia, had much of its northern landmass folded, sheared and compressed into the Himalayas. Tectonic extinction (Ebach 2003) will see areas of high biodiversity such as the Caribbean, the Mediterranean Sea and Wallacea vanish along with many organisms that currently live there in the near geological future (see Chap. 4).

Tectonics drives evolution and extinction through the twin processes of isolation leading to speciation and habitat destruction leading to loss of biodiversity. Without tectonics, our Earth's flora and fauna would be much more homogenous with low rates of both evolution and extinction, or perhaps may never have evolved life at all.

References

Capriolo M, Marzoli A, Aradi LE et al (2020) Deep CO2 in the end-triassic central atlantic magmatic province. Nat Commun 11:1670–1679

Ebach MC (2003) Tectonic extinction. Australas Sci 24:33–36

Ebach MC, Michaux B (2017) Establishing a framework for a natural area taxonomy. Acta Biotheor 65: 167–177

Ebach MC, Murphy DJ (2020) Carving up Australia's arid zone: A review of the bioregionalisation of the Eremaean and Eyrean biogeographic regions. Aust. J. Bot. 68:229–244. https://doi.org/10.1071/BT19077

Fergusson CL, Henderson RA (2015) Early Palaeozoic continental growth in the Tasmanides of Northeast Gondwana and its implications for Rodinia assembly and rifting. Gondwana Res 28:933–953

Flament N (2014) Linking plate tectonics and mantle flow to Earth's topography. Geology 42:927–928

Flament N, Gurnis M, Müller RD, Bower DJ, Husson L (2015) Influence of subduction history on South American topography. Earth Planet Sci Lett 430: 9–18

Large RR, Halpin JA, Lounejeva E, Danyushevsky LV, Maslennikov VV, Gregory D, Sack PJ, Haines PW, Long JA, Makoundi C, Stepanov AS (2015) Cycles of nutrient trace elements in the Phanerozoic ocean. Gondwana Res 28:1282–1293

Michaux B. 2010. Biogeology of Wallacea: geotectonic models, areas of endemism, and natural biogeographical units. Biol J Linn Soc 101: 193–212.

Myers N, Mittermeier RA, Mittermeier CG, Da Fonseca GA, Kent J (2000) Biodiversity hotspots for conservation priorities. Nature 403:853–858

Müller RD, Flament N, Matthews KJ, Williams SE, Gurnis M (2016) Formation of Australian continental margin highlands driven by plate–mantle interaction. Earth Planet Sci Lett 441: 60–70

Murphy DJ, Ebach MC, Miller JT, Laffan SW, Cassis G, Ung V et al (2019) Do phytogeographic patterns reveal biomes or biotic regions? Cladistics 35:654–670. https://doi.org/10.1111/cla.12381

Quigley MC, Clark D, Sandiford M (2010) Tectonic geomorphology of Australia. Geol Soc Lond Spec Publ 346: 243–65

Sandiford M (2007) The tilting continent: a new constraint on the dynamic topographic field from Australia. Earth Planet Sci Lett 261:152–163

Vaughan APM, Leat PT, Pankhurst RJ (2005) Terrane processes at the margins of Gondwana. Geol Soc Lond Spec Publ 246:1–21

Wignall P (2017) Mass Extinctions of Pangea (Jean Baptiste Lamarck Medal Lecture). In: Proceedings 19th EGU General Assembly, EGU2017, 23–28 April 2017, Vienna, p 2290

Contents

Chapter 1
Introduction to Neotectonics and Bioregionalisation

Abstract Neotectonics has undergone incredible development since Vladimir Obruchev proposed the term in 1948. With the discovery of mid-oceanic ridges, subduction, mantle flow, and dynamic topography, neotectonics has become the forefront of tectonic research. This chapter attempts to unite two fields, neotectonics and bioregionalisation, the latter being a result of the former.

1.1 Neotectonics: Dynamic Topography and Landscape Evolution

Neotectonics, a subdiscipline of tectonics, deals with Neogene intra-plate deformation due to the down and upwellings in the mantle that lead to dynamic topography and landscape development (e.g. formation of cratonic basins or alteration of drainage and erosion patterns) within recent geological time periods (i.e. since the Miocene). Although these processes operating on landscapes are presumed to have operated throughout Earth's geological history, only recent geomorphological features associated with neotectonics processes will still be preserved due to the ravages of erosion. The term neotectonics was first coined by Russian geologist Vladimir Obruchev (1863–1956), who proposed to "call the structures of the Earth's crust as neotectonics. Such structures had been created during the youngest movements of the crust, which had occurred at the end of the Tertiary Period as well as in the first half of the Quaternary Period". Obruchev did not support Wegner's continental drift hypothesis but rather considered "the movements (moving) of the Earth's crust, namely the processes of geotectonics, from the point of view of the pulsation hypothesis of [Mikhail Antonovich] Usov and [Walter Hermann] Bucher" (Obruchev 1948, p. 13, translation by E.V. Mavrodiev). Present-day neotectonics incorporates Obruchev's youngest structures as well as plate tectonic theory, in particular mantle flow due to slab subsidence or divergent margins.

1.1.1   The Geoid

In order to understand neotectonics, it is first necessary to introduce the concept of the geoid (Fig. 1.1). Imagine a snapshot of the entire surface of the Earth. In this two-dimensional picture, we can see the surface of the oceans without any influence from the tides, wind, or air pressure. The geoid, a hypothetical surface, which extends under the continents, is equivalent to a mean sea level. The only influences on the shape of the geoid are gravity and the rotation of the planet. Because the Earth's gravity varies due to the uneven distribution of mass and heterogeneity in the density of the underlying mantle, the geoid shows how differences in crustal thicknesses and irregularities in the upper mantle effect the elevation of oceans and continents with respect to the reference ellipsoid, which is another hypothetical surface representing the best approximation of the Earth's true shape. The reference ellipsoid assumes the Earth is homogeneous and, unlike the geoid, is smooth.

When we compare the geoid with the reference ellipsoid, we find minor and major discrepancies between their surfaces. In areas of high gravity, such as subduction zones where cold, dense lithosphere enters the upper mantle or anomalously dense rocks are concentrated, the surface of the geoid will subside. In areas of low gravity, formed by the upwards convection of hotter and therefore less dense mantle material or thickened low-density crust, it is raised. Comparing the geoid's undulation to the actual shape of the planet, namely, the reference ellipsoid, we find that gravity variations and rotation alone are responsible for significant areas of uplift and subsidence. Long wavelength variations in the geoid can generate large anomalies of ±100 m with wavelengths in the order of 10^3 kilometres.

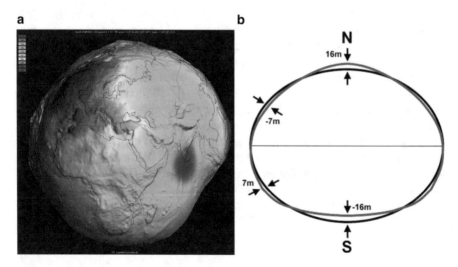

Fig. 1.1 (**a**) The geoid; (**b**) Geoid undulation relative to the reference ellipsoid. Undulation in false color, shaded relief and vertical exaggeration (10000 scale factor). Author: International Centre for Global Earth Models (ICGEM). Licensed under the Creative Commons Attribution 4.0 International license. (From Wikimedia commons https://commons.wikimedia.org/wiki/File:Geoid_undulation_10k_scale.jpg)

1.1.2 Dynamic Topography

As Ruby et al. (2017) noted, topographic features exist at various spatial scales. The sort of topography that is directly visible has been formed from an interaction between tectonics, crustal structures such as faults, and weathering processes leading to erosion. At larger spatial scales, topography is a result of an equilibrium between isostatic forces due to crustal and lithospheric mantle composition and thickness (and summarised by the geoid) and a dynamic component due to mantle convection. This dynamically sustained vertical deformation is called dynamic topography (Hager et al. 1985; Lithgow-Bertelloni and Gurnis 1997). The difference between long wavelength (10^3 km) topography and the geoid – also called residual topography – can be in the order of ±1 km and has been interpreted as a consequence of underlying upper mantle convection acting on the overlying lithosphere (Hoggard et al. 2016) and/or deeper mantle effects caused by, for example, fossil subduction slabs (Heine et al. 2008). In areas of mantle downwelling, the crust is depressed, and in areas of upwelling, the crust is elevated. Areas of upwelling are associated with spreading ridges, super plumes, and hotspots and areas of downwelling with active or fossil trenches (Fig. 1.2). The effects of density heterogeneity and dynamic topography are coupled and positively correlated. Gravity

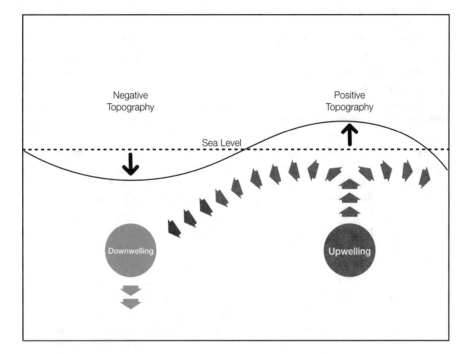

Fig. 1.2 A simplified example of upwelling and downwelling and its effects on dynamic topography

increases in trenches as cold, denser lithosphere enters the upper mantle and depresses the geoid while simultaneously causing downwards deflection of the overriding crust producing a negative dynamic topography. Gravity is decreased when lower-density mantle material rises causing the geoid to increase simultaneously with elevation of the overlying crust.

1.1.3 Landscape Development

The elevation or depression of land surfaces in stable continental areas, while less spectacular than the large vertical displacements possible through tectonic processes, are nevertheless important factors in the development of continental landscapes (Braun 2010) and ultimately, as we argue later, on the distribution and evolution of their biotas. Because cratonic areas are so flat, even a 100 m vertical displacement of the geoid will have significant effects on drainage patterns and the distribution of water in arid areas, and even quite modest tilting can cause extensive marine transgressions or regressions. Geomorphological effects are more pronounced when a dynamic component is also involved. For example, the formation of large epicontinental seas, long regarded as something of an enigma because they are formed on stable cratonic areas far from tectonically active regions, has been shown to be the result of tectonic tilt and/or mantle flow. Major epicontinental seas are known from the Cretaceous of North America (Mitrovika et al. 1989; Heine et al. 2008) and from the Cenozoic of Australia (Heine et al. 2010; Ruby et al. 2017), and both have been interpreted in terms of dynamic topography. In North America a large inland sea was formed in the west during the Cretaceous (Western Interior Seaway) that isolated western from eastern regions. The basin was estimated to have a maximum vertical displacement approaching 3 km and a maximum width of 1400 km. The formation of the Western Interior Seaway was a consequence of subduction along the Pacific coast of the continent during the Upper Cretaceous, which caused tilting and dynamic depression, and was followed by rebound and uplift at the start of the Cenozoic when either subduction ceased or the angle of the Benioff Zone steepened (Mitrovika et al. 1989). Inland basins also formed in Australia and have also been interpreted in terms of a continent wide tilt: in a west-east direction from the Late Jurassic to Early Cretaceous due to westwards subduction along the Pacific coast of East Gondwana and in a south-north direction in the Neogene as Australia interacted with a series of subduction zones along its northern and northeastern margins.

An associated phenomenon is the movement of palaeoshorelines across the margins of stable continents. Marine regressions and transgressions on the margins of tectonically stable cratons have traditionally been interpreted in terms of sea-level changes. However, recent work by DiCaprio et al. (2009) has shown that inundation patterns in Australia are not consistent with sea-level changes. They argued that detailed analyses of palaeoshoreline changes during the Cenozoic in the west, north, and northeast Australia are consistent with a tilting of the continent about a roughly northwest-southeast axis. This resulted in a 300 m drop of Australia's northern margin since the Eocene that they attribute to a downwards flexure as the continent approached the Indonesian and Melanesian subduction systems. By the end of the Miocene and into the Pliocene, subsidence rates intensified flooding an extensive and long-lived carbonate platform in the northeast, as coral growth was no longer able to keep up with subsidence rates (DiCaprio et al. 2010). The geodynamic model of DiCaprio et al. (2009) does not account for the shoreline changes along the southern coast of Australia, which required an additional 250 m of vertical movement to account for Eocene and Miocene transgressions. This additional downwards flexure (and subsequent rebound) was, according to DiCaprio et al. (2009), a result of Australia's northwards movement across a north-south oriented fossil subduction slab.

Dynamic topography has also been implicated as a significant influence on drainage patterns and basin geometry in southern Africa (Roberts and White 2010), southern South America (Guillaume et al. 2009), and Australia (Czarnota et al. 2014). In the example from Patagonia, reported by Guillaume et al. (2009) and based on accurate mapping of old river terraces, river flows reversed direction as subduction migrated northwards beneath Patagonia during the Miocene causing pre-Miocene drainage basins to be uplifted as the overlying crust responded to cessation of mantle downwelling. These diverse phenomena clearly show that neotectonics can have a profound influence on landscape development in otherwise stable cratonic regions and that an understanding of landscape development in continents requires interpretations that take into account neotectonics processes.

1.2 Bioregionalisation

The theoretical and methodological differences between palaeobiogeography and biogeography lie in the availability of distributional and phylogenetic data. In addition, modern or extant biogeographical analyses mainly focus on terrestrial taxa, whereas palaeobiogeographical studies are mostly marine (Lieberman 2000). These differences have divided biogeographical and palaeobiogeographic studies, with biogeographical methods making their way into palaeobiogeography via terrestrial studies (e.g. Poropat et al. 2016) and a few marine studies (Lieberman 2000) and palaeobiogeographic methods being adopted by biogeographers who are short on cladograms (e.g. Biotic Analysis, see Dowding and Ebach 2018). However, the exchange of methodologies has not been accompanied by theoretical interchange. For instance, palaeontologists who have adopted biogeographic methods have generally avoided discussing bioregionalisation, endemicity, and biotic areas, instead favouring single taxon histories. Biogeographers have adopted Biotic Analysis but have avoided taking stratigraphy into account (see Dowding and Ebach 2018). A way to remedy this asymmetric reciprocity is to introduce the concept of area taxonomy (Ebach and Michaux 2017) in order to make sense of biotic areas in terrestrial palaeobiogeography. Area taxonomy is a subdiscipline of comparative biogeography (Ebach and Michaux 2017) that attempts to create hierarchical bioregionalisations of natural biotic areas in order to understand the biotic history of continental areas. Recent examples of area taxonomy used include bioregionalisation of the Australian flora (González-Orozco et al. 2014; Ebach et al. 2015) and Devonian bioregionalisation (Dowding and Ebach 2016, 2018). By classifying assumed natural biotic areas into large regions and realms, we can reconstruct the history of biotas through time. Like taxa, biotic areas can become extinct, move, or change over time. More importantly, by testing these biotic areas we may empirically discover whether they are natural (sensu Ebach and Michaux 2017) or composite, that is, artificial. Artificial biotic areas, such as artificial taxa, tell us nothing about the history of a biota.

While natural biotic areas are vital to understanding the history of a biota, areas in general have been largely ignored by the palaeontological literature. Many eighteenth- and nineteenth-century naturalists working on extant taxa understood that in order to explain animal and plant distributions, a concept of area was needed, as biotic areas or regions were defined by the composition of their native taxa. Those taxa that were shared between regions could be said to migrate from one area to another. Very early on areas became central to organismal distribution over the Earth and up until the mid-twentieth century Sclater-Wallacean areas (Sclater 1858; Wallace 1876) were quintessential to any palaeobiogeographical study (see Arldt 1907), as it was the result of millions of years of taxic distribution. In pre-tectonic thought older distributions based on extinct fossil taxa were thought to have occurred via land bridges that were part of sunken atolls or continents (e.g. von Ihering 1897; Arldt 1907; Schuchert 1932) or by long-range, chance dispersal events (Darwin 1859). The new paleogeographic reconstructions of the tectonic revolution had not revised these pre-tectonic ideas: land bridges were replaced by large-scale dispersals (Darlington 1957), and discussions about bioregionalisations were ditched in favour of single taxon dispersal histories (Simpson 1977). Biogeography of extant taxa also experienced a similar plight with the abandonment of bioregionalisation as "static" in favour of population or single taxa studies (Mayr 1951). The discussion about extant natural biogeographic areas and the ability to test them did undergo a revival between the 1970s and 2000s (Rosen 1979; Nelson and Platnick 1981; Nelson and Ladiges 1996; Ebach et al. 2005); however, while the methods were in place, the phylogenetic and distributional data were sadly lacking. It was not until the late 2000s that large distributional databases (e.g. Atlas of Living Australia; Global Biodiversity Inventory Facility) and the greater availability of cladograms finally made large-scale multi-taxon bioregionalisation studies possible (Holt et al. 2013; Kreft and Jetz 2010). Now that palaeontology too has large databases (e.g. Palaeobiology Database; Fossilworks) and many more cladistic studies (Poropat et al. 2016; Lieberman and Karim 2010), palaeontologists are now in a position to test biotic area hypotheses that have been debated since Suess (1906), Arldt (1907), and Boucot et al. (1969), for example, whether regions such as the mid-Palaeozoic Old World and Malvinokaffric were natural.

1.3 Integrating Neotectonics and Bioregionalisation

Changes to the geoid and any dynamic component to topography are subtle and are most easily recovered in areas distant from tectonically active zones where they are swamped by the far larger effects of tectonics interactions. Because these effects also occur at long wavelengths (in the order of 10^3 km), neotectonics processes are important in continental areas. Bioregionalisation is also a phenomenon most closely associated with continental areas, so what effect might neotectonics have on bioregionalisation patterns?

The formation of large epicontinental seas clearly has the potential to split ancestral ranges in multiple lineages causing a single biota to evolve into two or more distinct biotas. This effect is clearly observed in the theropod dinosaur faunas of the Campian (Upper Cretaceous) of North America following the formation of the Western Interior Seaway (Brownstein 2018). The diverse theropod fauna of the Atlantic Coastal Plain was similar to that of western regions of North America during the mid-Cretaceous but by the Campian was distinct. The Upper Cretaceous theropod fauna of eastern North America was relictual – dominated by primitive clades that had survived in isolation on the Atlantic Coastal Plain – while that of western regions had independently evolved into more derived forms.

Significant effects on the distribution of biotas can also result from smaller-scale geomorphological changes. While drainage patterns are an important factor in erosion and deposition and hence overall geomorphology, in arid areas such as Australia, it is in the distribution of scarce water resources that has the most significant biological effect by controlling vegetation patterns. This example will be discussed more fully in Chapter 3, but the tilting of Australia to the north in the Neogene has directed river flows towards the north and away from inland regions causing northern areas to become wetter and inland regions drier. As the distribution of water resources changes, so do the distributions of sclerophyllic and wetter floras. Finally, as the example of the drowned coral reefs in northeast Australia discussed above illustrates, these quite subtle changes in topography can also bring about local extinction of whole biological communities.

References

Arldt T (1907) Paläogeographisches zum Stammbaum des Menschen. Z Morphol Anthropol 10:203–215

Boucot AJ, Johnson JG, Talent JA (1969) Early Devonian brachiopod zoogeography. Geol Soc Am Spec Pap 119:1–113

Braun J (2010) The many surface expressions of mantle dynamics. Nat Geosci 3:825–833

Brownstein CD (2018) The distinctive theropod assemblage of the Ellisdale site of New Jersey and its implications for North American dinosaur ecology and evolution during the cretaceous. J Paleontol 92:1115–1129

Czarnota K, Roberts GG, White NJ, Fishwick S (2014) Spatial and temporal patterns of Australian dynamic topography from river profile modeling. J Geophys Res Solid Earth 119:1384–1424

Darlington PJ (1957) Zoogeography. The geographic distribution of animals. John Wiley and Sons, New York

Darwin C (1859) On the origin of species. John Murray, London

DiCaprio L, Gurnis M, Müller RD (2009) Long-wavelength tilting of the Australian continent since the late cretaceous. Earth Planet Sci Lett 278:175–185

DiCaprio L, Müller RD, Gurnis M (2010) A dynamic process for drowning carbonate reefs on the northeastern Australian margin. Geology 38:11–14

Dowding EM, Ebach MC (2016) The Early Devonian palaeobiogeography of Eastern Australia. Palaeogeogr Palaeoclimatol Palaeoecol 444:39–47

Dowding EM, Ebach MC (2018) An interim global bioregionalization of Devonian areas. Palaeobiodiversity Palaeoenvironments 98:527–547

Ebach MC, Michaux B (2017) Establishing a framework for a natural area taxonomy. Acta Biotheor 65:167–177

Ebach MC, Humphries CJ, Newman RA, Williams DM, Walsh SA (2005) Assumption 2: opaque to intuition? J Biogeogr 32:781–787

Ebach MC, Murphy DL, González-Orozco CE, Miller JT (2015) A revised area taxonomy of phytogeographical regions within the Australian Bioregionalisation atlas. Phytotaxa 208:261–277

Fergusson CL, Henderson RA (2015) Early Palaeozoic continental growth in the Tasmanides of Northeast Gondwana and its implications for Rodinia assembly and rifting. Gondwana Res 28:933–953

González-Orozco CE, Ebach MC, Laffan S, Thornhill AH, Knerr NJ, Schmidt-Lebuhn AN et al (2014) Quantifying Phytogeographic regions of Australia using geospatial turnover in species composition. PLoS One 9(3):e92558. https://doi.org/10.1371/journal.pone.0092558

Guillaume B, Martinod J, Husson L, Roddaz M, Riquelme R (2009) Neogene uplift of central eastern Patagonia: dynamic response to active spreading ridge subduction? Tectonics 28. https://doi.org/10.1029/2008TC002324

Hager BH, Clayton RW, Richards MA, Comer RP, Dziewonski AM (1985) Lower mantle heterogeneity, dynamic topography and the geoid. Nature 313:541–545

Heine C, Müller RD, Steinberger B, Torsvik TH (2008) Subsidence in intracontinental basins due to dynamic topography. Phys Earth Planet Inter 171:252–264. https://doi.org/10.1016/j.pepi.2008.05.008

Heine C, Müller RD, Steinberger B, DiCaprio L (2010) Integrating deep earth dynamics in paleogeographic reconstructions of Australia. Tectonophsics 483:135–150

Hoggard MJ, White N, Al-Attar D (2016) Global dynamic topography observations reveal limited influence of large-scale mantle flow. Nat Geosci 9:456–463

Holt BG, Lessard J-P, Borregaard MK, Fritz SA, Araújo MB, Dimitrov D, Fabre P-H, Graham CH, Graves GR, Jønsson KA, Nogués-Bravo D, Wang Z, Whittaker RJ, Fjeldså J, Rahbek C (2013) An update of Wallace's zoogeographic regions of the world. Science 339:74–78

Kreft H, Jetz W (2010) A framework for delineating biogeographical regions based on species distributions. J Biogeogr 37:2029–2053.

Lieberman BS (2000) What is paleobiogeography? In: Paleobiogeography. Topics in Geobiology 16. Springer, Boston, MA, pp 1–3

Lieberman BS, Karim TS (2010) Tracing the trilobite tree from the root to the tips: A model marriage of fossils and phylogeny. Arthropod Struct Dev 39(2–3):111–123. https://doi.org/10.1016/j.asd.2009.10.004

Lithgow-Bertelloni C, Gurnis M (1997) Cezozoic subsidence and uplift of continents from time-varying dynamic topography. Geology 25:735–738

Mayr E (1951) Bearing of some biological data on geology. GSA Bull 62:537–546

Mitrovika JX, Beaumont C, Jarvis GT (1989) Tilting of continental interiors by the dynamical effects of subduction. Tectonics 8:1079–1094

Nelson G, Ladiges PY (1996) Paralogy in Cladistic biogeography and analysis of Paralogy-free subtrees. Am Mus Novit 3161:1–58

Nelson G, Platnick NI (1981) Systematics and biogeography; cladistics and Vicariance. Columbia University Press, New York

Obruchev VA (1948) Osnovnye cherty kinetiki i plastiki neotektonik. Izv. Akad. Nauk, Ser. Geol. 5:13–24

Poropat SF et al (2016) New Australian sauropods shed light on Cretaceous dinosaur palaeobiogeography. Sci Rep 6:34467. https://doi.org/10.1038/srep34467

Roberts GG, White N (2010) Estimating uplift rate histories from river profiles using African examples. J Geophys Res Solid Earth 115. https://doi.org/10.1029/2009JB006692

Rosen DE (1979) Fishes from the uplands and Intermontane basins of Guatemala: revisionary studies and comparative geography. Bull Am Mus Nat Hist 162:267–376

Ruby M, Brune S, Heine C, Davies DR, Williams SE, Müller RD (2017) Global patterns in Earth's dynamic topography since the Jurassic: the role of subducted slabs. Solid Earth 8:899–919

Schuchert C (1932) Gondwana land bridges. GSA Bull 43:875–916.

Sclater PL (1858) On the general geographical distribution of the members of the class Aves. J Proc Linn Soc 2:130–136

Simpson GG (1977) Too many lines; the limits of the oriental and Australian zoogeographic regions. Proc Am Philos Soc 121:107–120

Suess E (1906) The face of the earth, vol 2. Clarendon, Oxford

von Ihering H (1897) Os Molluscos dos terrenos terciarios da Patagonia. Revista do Museu Paulista 21:217–382

Wallace AR (1876) The geographical distribution of animals. Macmillan, London

Chapter 2
Traversing Terranes: The Australides

Abstract We discuss the Australides, the orogenic belt that covers eastern Australia, New Zealand, East and West Antarctica, the Cape region of Africa, and South America, and summarize its tectonic history since the Neoproterozoic. A biotic analysis using palaeodistributional data is used to determine relationships between areas within the Australides. We integrate the palaeobiogeographic (phylogenetic) and tectonic histories in order to establish the extent in space and time of any Weddellian Province, and undertake an analysis of phylogenetic data (cladograms) to determine whether area relationships are driven by tectonostratigraphic terranes or by cratons and cratonic basins.

2.1 Introduction

A number of names have been used to describe a Neoproterozoic to Late Mesozoic orogenic belt that occurred along the palaeo-Pacific and Iapetus margins and later the Pacific margin of Gondwana. Vaughan et al. (2005) introduced the informal name "Australides" to describe the orogenic belt found in eastern Australia, New Zealand, East and West Antarctica, the Cape region of Africa, and South America (Table 2.1). Cawood (2005) termed this belt the "Terra Australis orogeny" and named the latest Triassic/Permian phase as the "Gondwanides orogeny". The term "Tasmanides" described the Neoproterozoic to Carboniferous orogens of eastern Australia (Foster and Gray 2000; Glen 2005; Rosenbaum 2018). These names refer to different elements of a set of related events. The most restricted in both space and time is the Tasmanides orogen of eastern Australia. Both the Australides and Terra Australis orogens are Gondwana wide, but the Terra Australis orogen ends in the Triassic (230 Ma) with the final assembly of Pangaea, while the Australides orogen lasts until the breakup of Eastern Gondwana in the Cretaceous (100 Ma). Indeed, it might be argued that the Australides is ongoing in New Zealand and South America.

Electronic Supplementary Material The online version of this chapter (https://doi.org/10.1007/978-3-030-51773-1_2) contains supplementary material, which is available to authorized users.

Table 2.1 Major events of Gondwana tectonics (570 Ma–present) and proposed correlations

Australia	NZ	East Antarctica	West Antarctica	Southern Andes	Central Andes	Cape Province	
	Kaikoura orogeny			Andean orogeny	Andean orogeny		0 ma
Extension	Extension	Extension	Extension	Andean orogeny	Andean orogeny		100 ma
	Rangitata orogeny		Peninsula orogeny Assembly of MBL	Chonide orogeny	Andean orogeny Extension		200 ma
Gympie terrane Hunter Bowen orogeny	Brook Street terrane	Beacon Supergroup		Extension	Alleghenide orogeny	Cape orogeny	
Lachlan – New England orogeny	Western Province Tuhua orogeny		Older intrusives of the Amundsen Province (MBL) Ross Province (MBL) Eastern Domain (AP)	Cuyania/ Precordillera and Chilenia terranes South Patagonia			300 ma
							450 ma
Ross-Delamerian orogeny		Robertson Bay turbidites Bowers Terrane Wilson Mobile Belt	EWM	Pampean orogeny	MARA Terrane	Saldanian orogeny	570 ma

AP Antarctic Peninsula, EWM Ellsworth-Whitmore Mountains, MBL Marie Byrd Land

We will follow Vaughan et al. (2005) and use the name Australides for the series of long-lasting and extremely complex cycles of convergence, terrane accretion, and formation of new continental crust along the Gondwanan margin.

The Australides date from shortly after the establishment of a convergent margin along the palaeo-Pacific and Iapetus margins of the newly assembled Gondwana between 590 and 550 Ma (Harley et al. 2013). According to Cawood (2005), these orogenies occurred along a zone 18,000 km long and up to 1600 km wide in what was clearly a major global tectonic feature. There is no evidence that any of the orogenic events along the Gondwana margin were caused by collision of continental blocks that were a feature of the Atlantic and Tethyan Oceans, which exhibited typical Wilson cycles of opening and closing ocean basins (Wilson 1968; Burke and Dewey 1974). Rather, the Australides represented a Cordilleran-type accretionary orogen over most of its length (Vaughan and Pankhurst 2008). The complexity of tectonic events within the Australides orogenic belt is outlined in Table 2.1 and Fig. 2.1 and will be discussed in detail below.

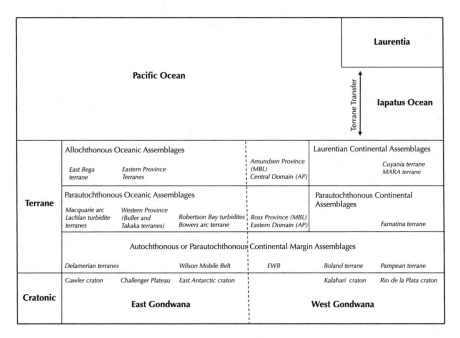

Fig. 2.1 Schematic summary of terranes involved in the Australides orogeny. Terranes in rows are equivalent in age and those in columns occupy equivalent geographical areas. See text for details

2.1.1 Early Phase (Neoproterozoic to Ordovician)

Evidence for an early convergent margin can be found in eastern Australia (Delamerian orogeny), Victoria Land in East Antarctica (Ross orogeny), southern Africa (Saldania orogeny), and southern South America (Pampean orogeny) (Table 2.1). The Delamerian orogeny of eastern Australia and Tasmania resulted in the inversion of Neoproterozoic marginal rift basins as a consequence of subduction that developed along the palaeo-Pacific margin of the North Australian and Gawler cratons between 530 and 500 Ma (Harley et al. 2013). The orogeny is dated at 514–490 Ma by Foden et al. (2006) and was initiated by an arc-continent collision in Tasmania and South Australia (Glen 2005). The formation of the Delamerian mountains and their subsequent erosion provided the siliciclastic sediments for thick turbidite sequences that were deposited outboard in back-arc basins, which were subsequently accreted during the Lachlan orogeny (Foster and Gray 2000; Glen et al. 2009). A similar pattern is recorded by contemporaneous events along the East Antarctic craton in northern Victoria Land (Harley et al. 2013). Here an autochthonous low P/high T metasedimentary rift sequence (Wilson "terrane") is intruded by the Granite Harbour Intrusives, a magmatic arc formed above a subduction zone and referred to as the Wilson Mobile Belt by Roland et al. (2004). The outboard Bowers Terrane, composed of weakly metamorphosed arc volcanics and sediments, was accreted to this margin during the Ross orogeny circa 500 Ma by westwards subduction beneath the East Antarctic craton (Capponi et al. 1999; Roland et al. 2004). According to Roland et al. (2004), the thick turbidite sequences of the Robertson Bay "terrane" are better regarded as an accretionary wedge formed between the Bowers Terrane and a new outboard subduction zone rather than as a separate terrane.

In the Cape region of southern Africa, Neoproterozoic rifting along the southwest margin of the Kalahari craton resulted in the deposition of clastics and limestones (Boland terrane) and eventually more distal turbidites as the basin widened and new oceanic crust formed (Swartland terrane). By circa 570 Ma, spreading had ceased, and the basins started to close resulting in a poorly developed collision event (Saldanian orogeny) dated to between 550 and 510 Ma (Rozendaal et al. 1999). Curtis (2001) described a similar sequence of deepening Lower to Middle Cambrian continental rift sediments in the Ellsworth-Whitmore Mountains (EWM), West Antarctica, which are conformably overlain by the latest Cambrian to Devonian Chrashsite Group. Curtis (2001) argued that despite the lack of any evidence of an Upper Cambrian orogenic event, the stratigraphic similarity with the Cape region of southern Africa supports an adjacent position of the two blocks. The Saldanian orogeny has been linked to the southern South American Pampean orogeny (Rapela et al. 2016; Casquet et al. 2018). The Pampean orogeny occurred during the collision of a Laurentian microcontinent (MARA terrane) with the Rio de la Plata craton – part of a west Gondwanan block that including the Kalahari craton – at 520 Ma (Rapalini 2005) or earlier (Casquet et al. 2018). The MARA terrane is presently exposed in the central Andes region, although it has been suggested that it extends into southern South America (Ramos 1988; Rapela et al. 2016). Eastern Pampean terrane sediments were derived from a Gondwana source and show typical Gondwanan detrital zircon age profiles. The western Pampean sediments show typical Laurentian zircon age profiles leading some authors to suggest they were deposited on MARA terrane basement (Rapela et al. 2016; Casquet et al. 2018).

2.1.2 Middle Phase (Ordovician to Carboniferous)

Following the Delamerian orogeny and associated uplift, thick turbidite sequences were deposited to the east in back-arc basins. These sedimentary units (Bendigo and Melbourne terranes) were accreted to the Australian Gondwana margin at about 445 Ma (Foster and Gray 2000; Glen 2005), which marks the start of the Lachlan orogeny. Whether this compressive phase was a result of westwards subduction under the Bendigo/Melbourne terranes (Glen 2005), oblique convergence (Glen et al. 2009) or eastwards subduction under the offshore Macquarie arc (Foster and Gray (2000) is not clear. A second phase of accretion occurred at about 390 Ma when the offshore Macquarie arc and associated basins approached and became sutured to the Gondwana margin. Glen (2005) suggested that as the western Bega terrane entered the collision zone, it caused the subduction zone to jump to the east. This second phase of compression may well have been due to oblique convergence, with Glen et al. (2009) arguing that the eastern Bega terrane had an Antarctic provenance and was moved north to its present position east of the Macquarie island arc. The final phase of compressive tectonics in the Ordovician to Carboniferous orogenic belt was the New England orogeny which resulted from west dipping subduction and the accretion of a continental arc/forearc basins and accretionary wedge to the Gondwana margin (Roberts and Engel 1987; Rosenbaum et al. 2012).

In New Zealand there was a similar pattern involving turbidite and arc terrane accretion to the Gondwanan margin during this period. The Western Province of New Zealand (Mortimer et al. 2014) consists of two Lower Palaeozoic terranes, the metasedimentary, predominantly turbidite Buller terrane and an island arc – the Takaka terrane. The Buller terrane has been correlated with similar-aged turbidite sequences in Australia, such as the Bendigo terrane and the Robertson Bay terrane of East Antarctica (Cooper 1989). The Buller and Takaka terranes had amalgamated by the Middle Devonian (Robertson et al. 2019) and were sutured to the margin of Gondwana during the Late Devonian to Carboniferous (370–330 Ma) Tuhua orogeny, which Gibson and Ireland (1996) regarded as the New Zealand equivalent of the Lachlan orogen. Vaughan and Pankhurst (2008) recognised three tectonostratigraphic belts spanning New Zealand and the West Antarctic regions of Marie Byrd Land (MBL) and the Antarctic Peninsula (AP). The innermost and oldest belt consists of the Western Province (NZ), Ross Province (MBL), and Eastern Domain (AP). Siddoway and Fanning (2009) considered that the Cambrian oceanic turbidites (Swanson Formation) of the Ross Province, which are intruded by Devonian to Carboniferous (375–340 Ma) subduction-related granites, are equivalent to the

Western Province rocks of New Zealand that are also intruded by the similar-aged Karamea Batholith and the arc and turbidite terranes of the Central Lachlan belt in eastern Australia. In Graham Land (AP), Eastern Domain rocks include Early Devonian to Early Carboniferous (393–325 Ma) granitoids and Late Carboniferous quartz-rich turbidites that were interpreted as an accretionary complex (Vaughan and Storey 2000). The Amundsen Province (MBL) is intruded by calc-alkaline granitoids of Ordovician to Silurian (450–420 Ma) age.

Despite uncertainties in terrane delineation and their accretion histories, a general consensus for the tectonic history of South America has emerged over recent decades. Following the Lower Palaeozoic Pampean orogeny discussed previously, a number of terranes were subsequently sutured to the Pampean margin of South America. The Cuyania terrane (Ramos 2004) or Precordillera terrane (Astini et al. 1995) is a Laurentian continental block (Ramos 2004; Rapalini 2005) amalgamated with the Pampean terrane during the Ordovician (Vaughan and Pankhurst 2008). The Famatina terrane was either an autochthonous Ordovician magmatic arc formed above an eastwards-directed subduction zone along the Pampean margin (Rapalini 2005) or an allochthonous intra-oceanic arc (Astini et al. 1995) sutured to the Gondwanan margin as a result of the Cuyania terrane's collision. The presence of ophiolites (Rapalini 2005) in a proposed Devonian suture zone further to the west has led to the proposal of a Chilenian terrane sutured to the western margin of the Cuyania terrane in the Late Devonian (Astini et al. 1995). Poor exposure of this proposed terrane means little is known about it; it may be an exotic Laurentian terrane, Gondwanan, or even a part of the Cuyania terrane (Rapalini 2005; Vaughan and Pankhurst 2008).

It has long been recognised that Patagonia was different from the rest of southern South America (Ramos 2008), although there is still considerable uncertainty about its boundaries, nature, and accretion history. Regarded initially as an allochthonous Gondwanan terrane accreted to the Rio de La Plata craton along the Rio Colorado Suture Zone during the Carboniferous to Triassic (Ramos 1988) or Permian (von Gosen 2003), more recent work by Pankhurst et al. (2006) suggested the suture zone is further to the south between the North Patagonian and Deseado massifs. Under this model, this South Patagonian terrane is Gondwanan and was sutured to the rest of Southern South America in the Permian (Domeier and Torsvik 2014).

2.1.3   Late Phase (Permian to Triassic)

Cawood (2005) referred to a widespread Gondwanan Permian-Triassic orogenic belt as the Gondwanide orogeny. In eastern Australia the Triassic Hunter-Bowen orogeny was identified by Hoy and Rosenbaum (2017) as part of the Gondwanides, which they interpreted as a result of global plate reorganisation during and after the final assembly of Pangaea. The Hunter-Bowen orogeny was episodic between the Middle Permian (circa 270 Ma) and Middle Triassic (235 Ma) and was associated with either the accretion of the Gympie terrane (as an exotic island arc) or its formation as a magmatic arc along the Gondwana margin above a west-dipping subduction zone. Hoy and Rosenbaum (2017) correlated the Gympie terrane with similar-aged arc terranes in New Caledonia (Teremba terrane) and New Zealand (Eastern Province Brook Street terrane). The Eastern Province also consists of obducted oceanic crust (Dunn Mountain terrane) and a number of turbidite terranes (Mortimer et al. 2014). These turbidite terranes were deposited in marginal (or more distal) basins during the Permian and Triassic, and at least some of them are exotic (Adams et al. 2009) with proposed depocentres adjacent to northeast Australia (e.g. Adams et al. 2007) or West Antarctica (e.g. Wandres et al. 2004a, b). This indicates that any subduction in the New Zealand sector during Gondwanide times may have been oblique or that part of the boundary was a transform. In East Antarctica sediment supply to the Permian and Triassic foreland sedimentary basins was, according to Paulsen et al. (2017), derived from newly uplifted outboard areas in West Antarctica. Evidence for subduction along the West Antarctic margin at this time is limited to the presence of Permo-Triassic arc volcanics in the Amundsen Province, Marie Byrd Land (Siddoway and Fanning 2009).

Deformation and uplift in the Cape Fold Belt of South Africa in the Permian (275–260 Ma) have been recognised locally as the Cape orogeny (Hansma et al. 2016). While at some distance from any subduction system, the Cape orogeny may be similar to that described in east Antarctica. In northern Peru the Alleghenide orogeny (305–260 Ma) resulted from the collision of the Tahuín terrane with the Gondwana margin, but further south there is little evidence of subduction-related tectonics (Ramos 2018). According to Ramos (2018), two phases of Gondwanide activity can be recognised here on the basis of metamorphism and intra-plate granite emplacement during crustal extension at circa 305 Ma and 260 Ma.

2.1.4   Post-Gondwanides (Jurassic–Recent)

Storey et al. (1987) introduced the term Peninsula orogeny to distinguish Jurassic and Early Cretaceous tectonics affecting the Antarctic Peninsula from the Permo-Triassic Gondwanide orogeny. The Peninsula orogeny (Palmer Land orogeny of Vaughan et al. 2002) was driven by subduction and resulted in the accretion of the Western and Central Domains to the Antarctic Peninsula margin. The Central Domain is composed of arc volcanics and may have been allochthonous (Vaughan and Storey 2000), formed in situ along the Antarctic Peninsula margin (Burton-Johnson and Riley 2015), or is composite (Ferraccioli et al. 2006). The Western Domain is an accretion complex of forearc basin turbidites (Le May Group) that Vaughan and Storey (2000) regarded as equivalents to other turbidite terranes in New Zealand (Eastern Province) and those of South America (Chonos terrane). Accretion probably occurred in the Early Cretaceous based on the timing of movement along the shear zone separating the western and central terranes from the parautochthonous Eastern Domain (Vaughan et al. 2002) and peak intrusive rates of the Peninsula Batholith (Leat et al. 1995). In New Zealand the Early Cretaceous

Rangitata orogeny was the result of amalgamation of the Eastern and the Western Provinces. The accretion of a number of terranes (Ramos 1988; Sepúlveda et al. 2008) and magmatism and metamorphism of Late Triassic turbidites (Hervé and Fanning 2001; Hervé et al. 2003) in South America during the Early Jurassic Chonide orogeny follow a broadly similar pattern to that seen in the Antarctic Peninsula and New Zealand but are a significantly earlier event than either the Rangitata or Peninsula orogenies. Further north in the central Andes, there was a prolonged period of arrested or slow subduction and back-arc extension (Charrier et al. 2014).

The Late Jurassic to mid-Cretaceous (circa 100 Ma) was a time of profound global tectonic change (Matthews et al. 2012) as the long-running compressive regime around the margin of southern Gondwana abruptly changed to one of extension, resulting in the eventual fragmentation of Gondwana. Rifting of the Pacific sector of Gondwana occurred between Australia and the Lord Howe Rise (Tasman and Coral Seas); between Australia and East Antarctica (Southern Ocean); between East and West Antarctica (West Antarctic Rift System); between the Campbell Plateau and Marie Byrd Land (Southern Ocean); and between South America and West Antarctica (Drake Passage). In Patagonia extension continued until the mid-Cretaceous when subduction resumed and the Andean orogeny started to effect southern South America at 91 Ma (Fildani et al. 2003). However, a change from compressive to extensional tectonics while widespread at 100 Ma was not universal. For example, in central South America, compression had started by the Early Jurassic (174 Ma) following a prolonged period of extension from the Late Permian (260 Ma). According to Charrier et al. (2014), this early period of the Andean orogeny (Early Jurassic to Early Cretaceous) was characterised by the resumption of subduction, deposition in back-arc basins, arc volcanism, and granitic magmatism. In the later period (Late Cretaceous to present), the orogeny was characterised by oblique subduction, inversion of the back-arc basins, eastwards migration of arc volcanism, and mountain building. O'Driscoll et al. (2012) have detailed the latest Cretaceous and Cenozoic development of the Andean orogeny in the central Andes, with initial mountain building dating from circa 70 Ma. This initial phase of the Andean orogeny was characterised by variable subduction rates along strike (70–50 Ma) followed by a general increase in rates between 50 Ma and 37 Ma. According to O'Driscoll et al. (2012), the strong compression experienced in the central Andes between 37 Ma and 25 Ma was the result of oblique subduction of the Nazca Plate and flat-slab subduction beneath the Amazonian Shield. In flat-slab subduction, the angle of subduction shallows and follows the base of the overriding craton. Post 25 Ma, subduction direction became orthogonal to the trench, subduction rates increased, and the angle of subduction steepened. Ramos and Ghiglione (2008) recognised a similar sequence of events in the southern part of Patagonia where strong uplift was experienced between 28 Ma and 26 Ma due to increased convergence rates and a change from oblique to orthogonal subduction. The main phase of Andean uplift in Patagonia was in the Miocene (circa 17 Ma). The Andean orogeny is unusual in the context of Gondwanan orogenies because it does not appear to have involved terrane accretion (Charrier et al. 2014).

Apart from subduction along the western margin of South America, the only remnant of the Neoproterozoic to Mesozoic subduction system of Pacific Gondwana is now found in the southwest Pacific. As Gondwana fragmented and the southern continents dispersed, Antarctica became surrounded by spreading centres, and the remaining subduction boundary migrated into the Pacific. At present the South Island, New Zealand, straddles the boundary between the Australian and Pacific plates. The Pacific plate is being subducted beneath the Australian plate to the north of New Zealand along the Tonga/Kermadec Trench, the Vityaz Arc, and along the north coast of New Guinea (where the boundary is much fragmented). To the south polarity is reversed and the Australian plate subducts beneath the Pacific plate. These two subduction regimes are connected by the Alpine Fault, a dextral transform boundary that runs through the South Island of New Zealand. The development of the present boundary can be traced to the late Oligocene (circa 24 Ma) when the Pacific/Australian plate boundary migrated into northern New Zealand eventually resulting in uplift and the rise of the Southern Alps and Seaward Kaikoura Range in the Plio-Pleistocene (5 Ma). This latest period of mountain building is known as the Kaikoura orogeny.

2.2 Biogeographic Analysis of the Australides

2.2.1 Bioregions and Tectonics

Figure 2.2 summarises the complex tectonic developments of southern Gondwana during the latest Palaeozoic and Mesozoic and shows the arrangements of cratons, terranes, and marginal basins along the Australides orogenic belt. Figure 2.2a shows the approximate spatial extent of the Australides orogenic belt prior to the breakup of southern Gondwana at 100 Ma. Figure 2.2b is a schematic representation of southern Gondwana showing the distribution of terranes, rift basins, and cratons post –100 Ma with letters referring to areas shown in Fig. 2.2a. The development of the boundaries between areas through time is shown in Fig. 2.2c with numbers referring to the sections shown in Fig. 2.2b. In Fig. 2.2c the term marginal basin encompasses any basin type formed along the margins of cratons, such as rift or back-arc basins, while the term terrane also includes the Late Cretaceous to Present Andean Orogeny that was not the result of terrane accretion.

Prolonged periods of mountain building, the formation and biotic dispersal (i.e. geodispersal) of terranes, and periods of extension would have had an evolutionary impact by increasing allopatric speciation rates as new biological barriers were formed; through increasing extinction rates as existing environments were destroyed or significantly modified; and by promoting range expansion into newly created or expanded habitats. These evolutionary effects would have been largely confined to terrane regions. The eroded landscapes and uniform environmental conditions found over large areas in cratonic regions, in contrast to the environments of orogenic belts, promote evolutionary stability. One can consider this a form of geo-

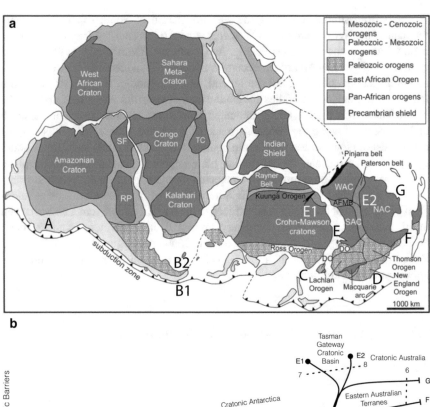

Fig. 2.2 (**a**) Geographic extent of the Australides orogeny. Letters are areas referred to in (**b**) changes in tectonic structure along the Australides orogenic belt and (**c**) Changes in tectonic structure in time and space

graphical punctuated equilibrium on a regional scale. The Australides are an ideal example to test the hypothesis that evolution is driven by tectonics. While a test of this hypothesis for the entire Australides orogen in time and space is beyond the scope of this book, a more restricted test of Mesozoic and Palaeogene distributions in cratonic and terrane provinces of southern Gondwana is feasible.

We suspect that marginal mountain chains, formed in response to terrane amalgamation or subduction, would isolate terrane from cratonic biotas to a greater or lesser degree. However, from the late Jurassic (~150 Ma) through to the mid-

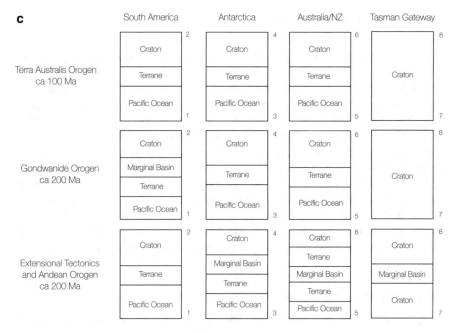

Fig. 2.2 (continued)

Cretaceous (100 Ma), South America, Antarctica, and Australasian terrane taxa would have been increasingly geographically isolated from their respective cratonic taxa due the complex system of rift basins that formed as Gondwana started to break up (Fig. 2.2). This complex geography inland of the Australide margin would have acted as a significant long-term biological barrier preventing taxa on terranes from mixing with taxa on cratons and resulting in terrane and cratonic biotas evolving independently. While biological barriers existed between cratons and terranes, barriers within terranes and cratons would have been fewer and more ephemeral. We would argue that within a craton or along terranes, there was far less biological isolation because there are few barriers to dispersal other than distance within cratonic regions, and active tectonics along subduction zones would continually disrupt any barriers either through terrane amalgamation or mountain building forming expansion corridors within or between terranes. If isolation between terrane taxa and cratonic taxa was evolutionarily significant, then a phylogenetic pattern would have emerged where terrane taxa are more closely related to other terrane taxa than they are to cratonic taxa. These processes, if the hypothesis of tectonics as a major driver of evolution is correct, should have led to provincialisation and the formation of bioregions within Gondwana such as a South American-Antarctica-Australasia provincialisation where cratons are biologically linked to other cratons and terranes to other terranes.

2.2.2 Weddellian Province

How might one test the hypothesis that tectonics drives evolution, namely, whether the Australides orogenic belt has influenced the terrestrial bioregionalisation of South America, Antarctica, and Australasia into biotic areas that are endemic to either terranes or cratons but not both? There is already indication from the literature that bioregionalisation had occurred in southern Gondwana. A Weddellian Province was proposed by Zinsmeister (1982) to account for similarities in the Late Cretaceous and Palaeogene gastropod faunas of southern South America, West Antarctica, New Zealand, and Australia. Subsequently, a similar pattern of provincialisation has been recognised in a diverse range of taxa including forest communities (Contreras et al. 2013; Vajda and Raine 2010; Wilf et al. 2009), mammals (Reguero et al. 2002; 2013 Tejedor et al. 2009), squamata (Martin and Fernández 2007), ammonites (Macellari 1987), and foraminifera (Huber 1992) and has also been proposed for elements of the modern fauna such as notothenioid fish (Near et al. 2015).

In order to find evidence to test the hypothesis that tectonics drives evolution, a comparative analysis will be employed that will:

1. Quantify terrestrial bioregionalisation relationships to establish the extent in space and time of any Weddellian Province (i.e. southern Gondwanan province) using the palaeodistributional data of fossil plants and animals and biotic area similarity analysis (Dowding et al. 2018).
2. Analyse the phylogenetic data of fossil taxa to find any evidence that contradicts a terrane- or craton-only relationship.
3. This model may only be considered valid once the palaeodistributional data and phylogenetic evidence do not contradict terrane-only and craton-only regions and relationships.

2.3 Comparative Analysis

2.3.1 Palaeodistributional Data and Biotic Analyses

Distributional data for plant (38), mammal (3), bird (4), fish (7), dinosaur (40), brachiopod (5), echinoid (13), gastropod (37), bivalve (10), and cephalopod (23) fossil taxa were retrieved from the Paleobiology Database (https://paleobiodb.org/#/) for the Triassic, Jurassic, Cretaceous, and Palaeogene (Table A.1). Genera or families (and some higher taxa) were used that occurred in at least two southern Gondwanan areas. Cratonic areas included Australia, South America, and East Antarctica, and terrane areas eastern Australia, western South America and Patagonia, West Antarctica, and New Zealand. A taxon-area matrix was constructed, and analyses of the complete data set and individual time slices (Dowding et al. 2018) were undertaken using the cladistics software TNT v1.5 (Goloboff et al., 2008), New Technology Search algorithm (1000 iterations, Implied Weighting K = 3) to find the most parsimonious cladograms. Trees were rooted using an all-zero outgroup.

2.3.2   Results of the Biotic Analyses

The results of the biotic analyses are shown in Fig. 2.3. Two shortest trees were produced when the whole data set was analysed, and the consensus tree is shown in Fig. 2.3a. Unsurprisingly, there was better resolution for the Cretaceous and Palaeogene periods because there were more data (see Table 2.2). Palaeogene areas form a clade that is sister area to a clade of all Cretaceous areas showing that there was a continuity between these two time periods. A Palaeogene Weddellian Province is clearly defined by the data, and furthermore all terrane areas within this province are more closely related to each other and form a clade that is sister area to cratonic

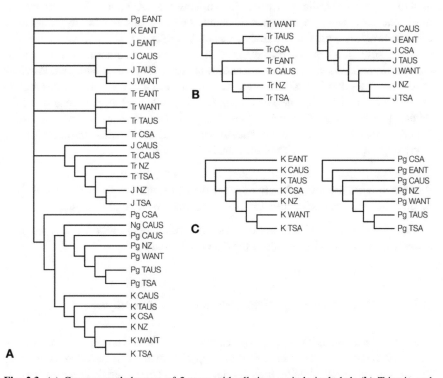

Fig. 2.3 (a) Consensus cladogram of 2 trees with all time periods included; (b) Triassic and Jurassic time slices analysed separately; (c) Cretaceous and Palaeogene time slices analysed separately. Abbreviations (Areas): *CSA* = Cratonic South America; *TSA* = Terrane South America; *EANT* = East Antarctica; *WANT* = West Antarctica; *CAUS* = Cratonic Australia; *TAUS* = Terrane Australia; *NZ* = New Zealand; (Periods) *Tr* = Triassic; *J* = Jurassic; *K* = Cretaceous; *Pg* = Palaeogene; *Ng* = Neogene

Table 2.2 Number of fossil species in the data set by area and age

	TSA	CSA	WANT	EANT	NZ	TAUS	CAUS	
Triassic	33	10	4	1	21	24	8	**101**
Jurassic	65	6	20	2	35	20	8	**156**
Cretaceous	92	63	82	1	69	47	33	**388**
Palaeogene	67	4	58	14	64	65	38	**310**
	257	**83**	**164**	**18**	**189**	**156**	**87**	

Australia. A Weddellian Province was also present in the Cretaceous with New Zealand, Western Antarctica, and terrane areas of South America forming a clade that is sister area to cratonic South America. The more basal position of Cretaceous terrane areas of Australia, resulting in their non-inclusion in the clade containing other terrane areas, is most probably in part an artefact of the relative paucity of Cretaceous fossil data for this area (Table 2.2). A further three clades are recovered in the analysis of the whole data set: a biochronological clade linking Jurassic and Triassic terrane areas of South America with those of New Zealand which are sister area to cratonic Australia; a Jurassic clade linking West Antarctica to Australian terrane areas as a sister area to cratonic South America; and a Triassic grouping of cratonic South America with terrane areas of Australia.

The results of a time slice analysis of the Triassic and Jurassic data, as well as Cretaceous and Palaeogene data, are shown in Figs. 2.3b, c. A Weddellian Province is recovered for the Jurassic with all terrane areas in a clade that is sister area to cratonic South America. The Triassic data are less resolved, in part because of the relative paucity of data (Table 2.2). One clade links New Zealand with terrane South America that is a sister area to cratonic Australia, and a second links terrane Australia to cratonic South America. The biotic analyses have shown:

1. The Weddellian Province originated by at least the Jurassic
2. That terranes are more closely related to other terranes
3. That the relationships within the terrane clades are variable through time

Whether this latter conclusion is a reflection of changing relationships amongst areas or lack of resolving power of the data is uncertain.

2.3.3 Towards an Empirical Biogeographic Analysis

Biotic similarity analysis lacks phylogenetic input and is, therefore, a preliminary approach that can be useful in suggesting hypotheses that can be tested using more rigorous methods based on biological relationships. These methods require phylogenies of taxa distributed in the areas of interest, which are then processed to produce consensus areagrams that summarise all the available information. Processing is required because individual areagrams (formed by substituting areas for taxa) are unlikely to be congruent with each other and the problem arises as to how best to deal with this incongruency. Apart from incompatible area relationships suggested by individual phylogenies, the major problems to address include the occurrence of widespread taxa producing MASTs (multiple areas on a single terminal branch), redundant areas (repeated occurrence of the same area on different branches on the areagram), and missing areas (Ebach et al. 2005; dos Santos 2011). Redundancy is usual in most areagrams and results in paralogous nodes that yield no information about area relationships and can obscure any cladistic signal (Nelson and Ladiges 1996). Using only non-paralogous nodes removes both the noise caused by redundant areas and the inherent bias of paralogy. There has also been vigorous debate about the best way of dealing with MASTs and missing taxa. For instance, Ebach et al. (2005) argued that the procedures for solving distribution patterns and resolving paralogy should be separated because the implementation of Assumption 2 (Nelson and Platnick 1981) may itself produce paralogous nodes and suggested a refinement of Assumption 2, termed the transparent method, for overcoming this problem. Processing cladograms produces MAST- and paralogy-free subtrees that can serve as inputs into biogeographic packages such as LisBeth (Zaragüeta-Bagils et al. 2012; http://infosyslab.fr/downloadlisbeth/LisBeth.exe), a programme that produces optimal trees from all the three-item statements employing an exhaustive branch and bound algorithm using compatibility analysis. An intersection tree is formed from all the three-item statements common to the optimal trees and is equivalent to a consensus tree.

Unfortunately, this analytical protocol cannot be fully implemented because of the lack of available cladograms. Cladograms of extant taxa cannot be used because of the absence of species in Antarctica and so are reliant on cladograms of fossil taxa. There are still surprisingly few cladistic analyses available for fossil taxa and even fewer covering taxa that are found in Weddellian areas (e.g., Lieberman and Karim 2010). Those available in the literature are shown in Table 2.3. Three-item statements of hypothetical area relationships involving cratons, terranes, and marginal basins (Table 2.4) produce 18 arrangements, 15 of which are informative and 7 of which support a Weddellian Province, that is, they show that terrane areas are more closely related to each other than they are to cratonic areas and vice versa.

Table 2.3 Fossil cladograms available in the literature, which include Jurassic-Late Cretaceous taxa from the Australides

Period/epoch	Cladograms (c,(t,t)), (t,(c,c)) or ((c,c),(t,t))	References
Jurassic-Late Cretaceous	Non-coelurosaurian tetanurans (theropods)	Smith et al. (2008)
Late Cretaceous	Titanosauria + Lithostrotia	Poropat et al. (2016)
Cretaceous	Ankylosauridae	Thompson et al. (2011)
Palaeocene-Oligocene	Dasyuromorphia	Kealy and Beck (2017)
Late Jurassic-Late Cretaceous	Dromaeosauridae	Choiniere et al. (2014)

Table 2.4 The seven relationships that support a Weddellian Province

	Informative	Uninformative
		(C(C,C))
		(T(T,T))
		(MB(MB,MB))
Support Weddellian Province	(C(T,T))	
	(C(MB,T))	
	(T(C,C))	
	(T(T,MB))	
	(MB(MB,T)) (MB(T,T)) (T(MB,MB))	
Neutral	(T(MB,C)) (MB(MB,C))	
	(C(MB,MB)) (C(C,MB)) MB(C,C)	
Do not support Weddellian Province	(T(T,C)) (C(C,T))	
	(MB(T,C))	

T terranes, *C* cratons, and *MB* marginal basins

Of the five cladograms available, all five resolve to three area statements which support an Australides relationship in which terrane areas are more closely related to each other than they are to cratons (relationships 1 and 3). While not a full analysis, these results are consistent with a Mesozoic Weddellian Province and provide some phylogenetic-based support for the result of the biotic analyses described earlier.

2.4 Tectonics as a Driver of Biodiversity

A Weddellian Province was the result of the independent evolution of a biota restricted to terrane fragments and marginal basins involved in the later stages of the Australides orogeny. There is no reason not to suppose that similar patterns of regionalisation occurred throughout the duration of this prolonged, episodic period of mountain building. Indeed, changes in the Lower Palaeozoic benthic marine fauna of the Precordilleran (Cuyanian) terrane illustrate precisely how rift, drift, and amalgamation of a terrane result in evolutionary change and increase in overall biodiversity. Benedetto (2004) reviewed the palaeontological evidence for the rift/drift/amalgamation history of this terrane and documented the change from a typical Cambrian Laurentian trilobite fauna to increasingly isolated and endemic Middle Ordovician benthic biotas and their eventual replacement by a Gondwanan biota by the end of the Ordovician.

Rifting of the Precordillerian microcontinent from the Ouachita embayment, Laurentia, during the Early Cambrian transported a trilobite fauna typical of the Laurentian "olenellid realm" into the Iapetus Ocean and remained strongly Laurentian until the end of the Cambrian when this fauna still showed strong affinity with a coeval Appalachian biota. The drift phase during the Early Ordovician was characterised by increasing isolation as the terrane evolved a biota characteristic of other Iapetus microcontinents (Avalonia) and intra-oceanic island arcs ("Celtic element") and a distinctive endemic element in the trilobite (60%) and brachiopod (25%) faunas. Endemism rates increased until by the Middle Ordovician 45% of sponges and 60% of benthic ostracods and trilobites were restricted to this terrane. The Middle Ordovician was also the time when the first elements of a Gondwanan fauna were recorded with approximately 10% of ostracods and 25% of trilobites showing a Gondwanan affinity and interpreted by Benedetto (2004) as a consequence of the terrane's approach to the Gondwanan margin. By the Late Ordovician, the overall Gondwanan element had increased to 65% in brachiopods, 20% in bivalves, 10% in ostracods, and 25% in trilobites. Interestingly, and despite the increase in Gondwanan influence, biological isolation remained high with 70% of bivalves, 50% of ostracods, and 40% of trilobites endemic to the terrane. Amalgamation of the Precordilleran terrane was complete by the beginning of the Silurian.

The biological history of this terrane was characterised by a progressive change from one fauna through local extinction and the independent evolution of taxa during the rift/drift phase to the eventual replacement by a new biota as Gondwanan taxa expanded their ranges during amalgamation. The result was the evolution of novel species and an increase in overall biodiversity that would not have occurred if the Precordilleran terrane had not rifted from Laurentia, been geodispersed across the Iapetus Ocean and sutured to Gondwana. And while this is a well-documented example it is not unique, and other examples of the independent evolution of biotas and their eventual transformation driven by tectonics are also known, for example, Chinese and Southeast Asian terranes. The Early Devonian (pre-rift) biotas of these terranes were Gondwanan but by the Carboniferous (drift phase) had evolved into a distinctive Cathaysian biota, which was replaced with a Eurasian biota in the Middle Jurassic (accretion phase) (Metcalfe 2013, Fig. 2.1). Tectonics does indeed seem to drive evolution.

References

Adams CJ, Campbell HJ, Griffin WR (2007) Provenance comparisons of Permian to Jurassic tectonostratigraphic terranes in New Zealand: perspectives from detrital zircon age patterns. Geol Mag 144:701–729

Adams CJ, Cluzel D, Griffin WL (2009) Detrital zircon ages and geochemistry of sedimentary rocks in basement Mesozoic terranes and their cover rocks in New Caledonia, and provenances at the Eastern Gondwanaland margin. Aust J Earth Sci 56:1023–1047

Astini RA, Benedetto JL, Vaccari NE (1995) The early Paleozoic evolution of the Argentine Precordillera as a Laurentian rifted, drifted, and collided terrane: A geodynamic model. Geological Society of America Bulletin 107:253–273

Benedetto JL (2004) The Allochthony of the Argentine Precordillera ten years later (1993–2003): a new Paleobiogeographic test of the microcontinental model. Gondwana Res 7:1027–1039

Burke K, Dewey JF (1974) Hot spots and continental breakup: implications for collisional orogeny. Geology 2:57–60

Burton-Johnson A, Riley TR (2015) Autochthonous v. accreted terrane development of continental margins: a revised *in situ* tectonic history of the Antarctic Peninsula. J Geol Soc 172:822–835

Capponi G, Crispini L, Meccheri M (1999) Structural history and tectonic evolution of the boundary between the Wilson and Bowers terranes, Lanterman Range, northern Victoria Land, Antarctica. Tectonophysics 312:249–266

Casquet C, Dahlquist JA, Verdecchia SO, Baldo EG, Galindo C, Rapela CW, Pankhurst RJ, Morales MM, Murra JA, Fanning CM (2018) Review of the Cambrian Pampean orogeny of Argentina; a displaced orogen formerly attached to the Saldania Belt of South Africa? Earth Science Reviews 177:209–225

Cawood PA (2005) Terra Australis Orogen: Rodinia breakup and development of the Pacific and Iapetus margins of Gondwana during the Neoproterozoic and Paleozoic. Earth Sci Rev 69:249–279

Charrier R, Ramos VA, Tapia F, Sagripanti L (2014) Tectono-stratigraphic evolution of the Andean Orogen between 31 and 37oS (Chile and Western Argentina). In: Sepúlveda SA, Giambiagi LB, Moreiras SM, Pinto L, Tunik M, Hoke GD, Farías M (eds) Geodynamic processes in the Andes of Central Chile and Argentina, Special publications, vol 399. Geological Society, London, pp 13–61

Choiniere JN, Clark JM, Norell MA, Xu X (2014) Cranial osteology of *Haplocheirus sollers* Choiniere et al., 2010 (Theropoda: Alvarezsauroidea). American Museum Novitiates 3816, 44pp

Cooper RA (1989) Early Paleozoic terranes of New Zealand. Journal of the Royal Society of New Zealand 19:73–112

Contreras L, Pross J, Bijl PK, Koutsodendris A, Raine JI, van de Schootbrugge B, Brinkhuis H (2013) Early to Middle Eocene vegetation dynamics at the Wilkes Land Margin (Antarctica). Rev Palaeobot Palynol 197:119–142

Curtis ML (2001) Tectonic history of the Ellsworth Mountains, West Antarctica: Reconciling a Gondwana enigma. GSA Bulletin 113:939–958

Domeier M, Torsvik TH (2014) Plate tectonics in the late Paleozoic. Geosci Front 5:303 –350

Dowding EM, Ebach MC, Mavrodiev EV (2018) Temporal area approach for distributional data in biogeography. Cladistics 35:435–445

Ebach MC, Humphries CJ, Newman RA, Williams DM, Walsh SA (2005) Assumption 2: opaque to intuition? J Biogeogr 32:781–787

Ferraccioli F, Jones PC, Vaughan APM, Leat PT (2006) New aerogeophysical view of the Antarctic Peninsula: More pieces, less puzzle. Geophys Res Lett 33. https://doi.org/10.1029/2005GL024636

Fildani A, Cope TD, Graham SA, Wooden JL (2003) Initiation of the Magallanes foreland basin: timing of the southernmost Patagonian Andes orogeny revised by detrital zircon provenance analysis. Geology 31:1081–1084

Foden J, Elburg MA, Dougherty-Page J, Burtt A (2006) The timing and duration of the Delamerian orogeny: correlation with the Ross Orogen and implications for Gondwana Assembly. J Geol 114:189–210

Foster DA, Gray DR (2000) Evolution and structure of the Lachlan Fold Belt (Orogen) of Eastern Australia. Annu Rev Earth Planet Sci 28:47–80

Gibson GM, Ireland TR (1996) Extension of Delamerian (Ross) orogen into western New Zealand: Evidence from zircon ages and implications for crustal growth along the Pacific margin of Gondwana. Geology 24:1087–1090

Glen RA (2005) The Tasmanides of eastern Australia. In: Vaughan APM, Leat PT, Pankhurst RJ (eds) Terrane processes at the margins of Gondwana, Geological society special publications, vol 246. Geological Society, London, pp 23–96

Glen RA, Percival IG, Quinn CD (2009) Ordovician continental margin terranes in the Lachlan Orogen, Australia: implications for tectonics in an accretionary orogen along the East Gondwana margin. Tectonics 28. https://doi.org/10.1029/2009TC002446

Goloboff PA, Farris JS, Nixon KC (2008) TNT, a free program for phylogenetic analysis. Cladistics 24:774–786

Gosen von W (2003) Thrust tectonics in the North Patagonian Massif (Argentina): Implications for a Patagonia plate. Tectonics 22. https://doi.org/10.1029/2001TC901039

Hansma J, Tohver E, Schrank C, Jourdan F, Adams D (2016) The timing of the Cape Orogeny: $^{40}Ar/^{39}Ar$ age constraints on deformation and cooling of the Cape Fold Belt, South Africa. Gondwana Research 32:122–137

Harley SL, Fitzsimons ICW, Zhao Y (2013) Antarctica and supercontinent evolution: historical perspectives, recent advances and unresolved issues. In: Harley SL, Fitzsimons ICW, Zhao Y (eds) Antarctica and supercontinent evolution, Geological society special publication, vol 383. Geological Society, London, pp 1–34

Hervé F, Fanning CM (2001) Late Triassic detrital zircons in meta-turbidites of the Chonos metamorphic complex, southern Chile. Rev Geol Chile 28:91–104

Hervé F, Fanning CM, Pankhurst RJ (2003) Detrital zircon age patterns and provenance of the metamorphic complexes of southern Chile. J S Am Earth Sci 16:107–123

Hoy D, Rosenbaum G (2017) Episodic behavior of Gondwanide deformation in eastern Australia: insights from the Gympie Terrane. Tectonics 36. https://doi.org/10.1002/2017TC004491

Huber BT (1992) Paleobiogeography of Campanian-Maastrichtian foraminifera in the southern high latitudes. Palaeogeogr Palaeoclimatol Palaeoecol 92:325–360

Kealy S, Beck R (2017) Total evidence phylogeny and evolutionary timescale for Australian faunivorous marsupials (Dasyuromorphia). BMC Evolutionary Biology 17:240–263. https://doi.org/10.1186/s12862-017-1090-0

Leat PT, Scarrow JH, Millar IL (1995) On the Antarctic Peninsula batholith. Geol Mag 132:399–412

Lieberman BS, Karim TS (2010) Tracing the trilobite tree from the root to the tips: A model marriage of fossils and phylogeny. Arthropod Struct Dev 39(2–3):111–123. https://doi.org/10.1016/j.asd.2009.10.004

Macellari CE (1987) Progressive endemism in the Late Cretaceous ammonite family Kossmaticeratidae and the breakup of Gondwana. In: McKenzie GD (ed) Gondwana six, sedimentology, and paleontology, Geophysical monographs, vol 41. American Geophysical Union, Washington, DC, pp 85–92

Martin JE, Fernández M (2007) The synonymy of the late cretaceous mosasaur (Squamata) genus *Lakumasaurus* from Antarctica with *Taniwhasaurus* from New Zealand and its bearing upon faunal similarity within the Weddellian Province. Geol J 42:203–211

Matthews KJ, Seton M, Müller RD (2012) A global-scale plate reorganisation event at 105-100 ma. Earth Planet Sci Lett 355-356:283–298

Metcalfe I (2013) Gondwana dispersion and Asian accretion: tectonic and palaeogeographic evolution of eastern Tethys. J Asian Earth Sci 66:1–33

Mortimer N, Rattenbury MS, King PR, Bland KJ, Barrell DJA, Bache F, Begg JG, Campbell HJ, Cox SC, Crampton JS et al (2014) High-level stratigraphic scheme for New Zealand rocks. N Z J Geol Geophys 57:402–419

Near TJ, Dornburg AD, Harrington RC et al (2015) Identification of the notothenioid sister lineage illuminates the biogeographic history of an Antarctic adaptive radiation. BMC Evol Biol 15:109. https://doi.org/10.1186/s12862-015-0362-9

Nelson G, Platnick NI (1981) Systematics and biogeography: cladistics and vicariance. Columbia University Press, New York

Nelson G, Ladiges PY (1996) Paralogy in cladistic biogeography and analysis of paralogy-free subtrees. Am Mus Novit 3167:1–58

O'Driscoll LJ, Richards MA, Humphreys ED (2012) Nazca-South America interactions and the late Eocene-late Oligocene flat-slab episode in the Central Andes. Tectonics 31. https://doi.org/10.1029/2011TC003036

Pankhurst RJ, Rapela CW, Fanning CM, Márquez M (2006) Gondwanide continental collision and the origin of Patagonia. Earth Sci Rev 76:235–257

Paulsen T, Deering C, Sliwinski J, Valencia V, Bachmann O, Guillong M (2017) Detrital zircon ages and trace element compositions of Permian-Triassic foreland basin strata of the Gondwanide orogen, Antarctica. Geosphere 13:2085–2093

Poropat SF et al (2016) New Australian sauropods shed light on Cretaceous dinosaur palaeobiogeography. Sci Rep 6(34467):2016. https://doi.org/10.1038/srep34467

Ramos VA (1988) Tectonics of the late Proterozoic–Early Paleozoic: a collisional history of Southern South America. Episodes 11:168–174

Ramos VA (2004) Cuyania, an Exotic Block to Gondwana: Review of a Historical Success and the Present Problems. Gondwana Research 7:1009–1026

Ramos VA (2008) Patagonia: A paleozoic continent adrift? Journal of South American Earth Sciences 26:235–251

Ramos VA (2018) Tectonic evolution of the Central Andes: from terrane accretion to crustal delamination. In: Zamora G, McClay KM, Ramos VA (eds) Petroleum basins and hydrocarbon potential of the Andes of Peru and Bolivia, AAPG Memoir, vol 117, pp 1–34

Ramos VA, Ghiglione MC (2008) Tectonic evolution of the patagonian andes. Dev Quat Sci 11:57–71

Rapalini AE (2005) The accretionary history of southern South America from the latest Proterozoic to the Late Palaeozoic: Some palaeomagnetic constraints. Geol Soc Spec Pub 246:305–328

Rapela C.W., Verdecchia S.O., Casquet C, Pankhurst R.J., Baldo E.G., Galindo C, Murra J.A., Dahlquist J.A., Fanninge C.M. 2016. Identifying Laurentian and SW Gondwana sources in the Neoproterozoic to Early Paleozoic metasedimentary rocks of the Sierras Pampeanas: Paleogeographic and tectonic implications. Gondwana Research 32:193–212

Reguero MA, Marenssi SA, Santillana SN (2002) Antarctic Peninsula and South America (Patagonia) Paleogene terrestrial faunas and environments: biogeographic relationships. Palaeogeogr Palaeoclimatol Palaeoecol 179:189–210

Reguero M, Goin F, Hospitaleche CA, Dutra T, Marenssi S (2013) Late Cretaceous/Paleogene West Antarctica Terrestrial Biota and its Intercontinental Affinities. SpringerBriefs in Earth System Sciences, https://doi.org/10.1007/978-94-007-5491-1_1

Roberts J, Engel BA (1987) Depositional and tectonic history of the southern New England Orogen. Australian Journal of Earth Sciences 34:1–20

Robertson AHF, Campbell HJ, Johnstone MR, Mortimer N (2019) Introduction to Paleozoic–Mesozoic geology of South Island, New Zealand: subduction-related processes adjacent to SE Gondwana. In: Robertson A.H.F. (ed.) Paleozoic–Mesozoic Geology of South Island, New Zealand: Subduction-related Processes Adjacent to SE Gondwana. Geological Society, London, Memoirs 491–14 https://doi.org/10.1144/M49.7

Roland NW, Läuffer AL, Rossetti F (2004) Revision of the Terrane Model of Northern Victoria Land (Antarctica). Terra Antarctica 11:55–65

Rosenbaum G (2018) The Tasmanides: phanerozoic tectonic evolution of Eastern Australia. Annu Rev Earth Planet Sci 46:291–325

Rosenbaum G, Li P, Rubatto D (2012) The contorted New England orogen (eastern Australia): new evidence from U-Pb geochronology of early Permian granitoids. Tectonics 31:TC1006. https://doi.org/10.1029/2011TC002960

Rozendaal A, Gresse PG, Scheepers R, Le Roux JP (1999) Neoproterozoic to Early Cambrian Crustal Evolution of the Pan-African Saldania Belt, South Africa. Precambrian Research 97:303–323

dos Santos CMD (2011) On the role of assumptions in cladistic biogeographical analyses. Papéis Avulsos de Zoologia 51. https://doi.org/10.1590/S0031-10492011001900001

Sepúlveda FA, Hervé F, Calderón M, Lacassie JP (2008) Petrological and geochemical characteristics of metamorphic and igneous units from the allochthonous Madre de Dios Terrane,
Southern Chile. Gondwana Res 13:238–249

Siddoway CS, Fanning CM (2009) Paleozoic tectonism on the East Gondwana margin: evidence from SHRIMP U–Pb zircon geochronology of a migmatite–granite complex in West
Antarctica. Tectonophysics 477:262–277

Smith ND, Makovicky PJ, Agnolin FL, Ezcurra MD, Pais DF, Salisbury SW (2008) A Megaraptor-
like theropod (Dinosauria: Tetanurae) in Australia: support for faunal exchange across
eastern and western Gondwana in the Mid-Cretaceous. Proceedings of the Royal Society
275:2085–2093

Storey BC, Thomson MR, Meneilly AW (1987) The Gondwanian orogeny within the Antarctic
peninsula: a discussion. In: Gondwana Six: Structure, Tectonics, and Geophysics, vol 40.
American Geophysical Union, Washington, DC, pp 191–198

Tejedor MF, Goin FJ, Gelfo JN, López GM, Bond M, Carlini AA, Scillato-Yan GJ, Woodburne
MO, Chornogubsky L, Aragón E, Reguero MA, Czaplewski NJ, Vincon S, Martin GM, Ciancio
MR (2009). New early Eocene mammalian fauna from western Patagonia, Argentina. American
Museum Novitiates 3638:1–43

Thompson RS, Parish JC, Maidment SCR, Barrett PM (2011) Phylogeny of the ankylosaurian
dinosaurs (Ornithischia: Thyreophora). Journal of Systematic Palaeontology 10(2):301–312.
https://doi.org/10.1080/14772019.2011.569091

Vajda V, Raine JI (2010) A palynological investigation of plesiosaur-bearing rocks from the upper
Cretaceous Tahora formation, Mangahouanga, New Zealand. Alcheringa 34:359–374

Vaughan APM, Pankhurst RJ (2008) Tectonic overview of the West Gondwana margin. Gondwana
Res 13(2):150–162

Vaughan APM, Storey BC (2000) The eastern Palmer Land shear zone: a new terrane accretion model for the Mesozoic development of the Antarctic Peninsula. J Geol Soc Lond
157:1243–1256

Vaughan APM, Kelley SP, Storey BC (2002) Mid-Cretaceous ductile deformation on the Eastern
Palmer Land Shear Zone, Antarctica, and the implications for timing of Mesozoic terrane collision. Geol Mag 139:465–471

Vaughan APM, Leat PT, Pankhurst RJ (2005) Terrane processes at the margins of Gondwana:
introduction. In: Vaughan APM, Leat PT, Pankhurst RJ (eds) Terrane processes at the margins
of Gondwana, Geological society special publications, vol 246. Geological Society, London,
pp 1–21

Wandres A.M., Bradshaw J.D., Weaver S., Maas R., Ireland T.R, Eby N. 2004a. Provenance analysis using conglomerate clast lithologies: a case study from the Pahau terrane of New Zealand.
Sedimentology 167: 57–89

Wandres A.M., Bradshaw J.D., Weaver S., Maas R., Ireland T.R, Eby N. 2004b. Provenance of the
sedimentary Rakaia sub-terrane, South Island, New Zealand: the use of igneous clast compositions to define the source. Sedimentology 168: 193–226

Wilf P, Little SA, Iglesias A, Zamola M, Gandolfo MA, Cúneo JKR (2009) Papuacedrus
(Cupressaceae) in Eocene Patagonia: a new fossil link to Australasian Rainforests. Am J Bot
96:2031–2047

Wilson JT (1968) Static or mobile earth: the current scientific revolution. Proc Am Philos Soc
112:309–320

Zaragüeta-Bagils R, Ung V, Grand A, Vignes-Lebbe R, Cao N, Ducasse J (2012) LisBeth: new
cladistics for phylogenetics and biogeography. Comptes Rendus Palevol 11:563–566

Zinsmeister WJ (1982) Late cretaceous – early tertiary molluscan biogeography of the southern
Circum-Pacific. J Paleontol 56:84–102

Chapter 3
Neotectonics and Australian Biogeography

Abstract Australia is the flattest continent on Earth and has a wide range of different landforms, making it an ideal place to investigate the impact of neotectonics (continental tilting and dynamic topography) on bioregionalisation. It is highly likely that continental tilting and dynamic uplift together have driven the biogeography of Australia since the Palaeogene.

3.1 Australia: The Titling Continent and the Birth of Biomes

The current narrative that the evolution of Neogene Australian biogeography, particularly its biomes, is a result of Australia's slowly drifting north into a new climate zone is too simplistic and needs to be recast in regards to biotectonics. Recent advances in geophysics and geomorphology have shown that dynamic topography due to intra-plate stresses has tilted the Australian continent along a NW-SE axis. This tilt, and the dynamic topographic changes that it has created, has had a greater effect on bedrock erosion than climate (Quigley et al. 2010). Bedrock erosion, due to changes in topography, is a result of weathering and later erosion due to drainage. The effects of drainage within the Australian continent are two-fold. Firstly, drainage directs water towards areas lower in altitude, thereby creating watersheds that conserve temperate and aseasonally wet biomes and create new monsoonal biomes. Without such watersheds, areas on the northern and eastern seaboards would be drier, possibly arid. Secondly, changes in topography have created drier regions such as an arid plateau in Western Australia with a reduced water catchment area and a lower arid basin between two watersheds in eastern Australia. While decreasing or increasing rainfall clearly has a general effect on vegetation (and thus biotic) patterns, the way in which that water is distributed has a direct influence on what grows where. It would be more accurate to attribute the evolution of Australia's biomes to a tilting continent rather than to a continent slowly moving north into a drier climate.

3.1.1 Introduction to Australia's Biomes

For over 150 years, climate has been considered the driving force behind Australian bioregionalisation. In 1889 Ralph Tate mapped three of Australia's endemic flora onto a "Rain Map of Australia" using average rainfall (Fig. 3.1). The Euronotian and Eremaen regions were proposed using the "line marking the 20-inch rainfall [...] which embraces the distinctive features botanically and physically of the tract exterior to the granite table land" (Tate 1889, p. 316). The autochthonous (Southwestern) region was similarly proposed, although the 10–25 in. and 25–50 in. lines "are narrowly separated". Tate's map was adopted in 1896 by Spencer for the fauna divisions of the Australian region using "the present rainfall limit of 25-50 in. per annum" (Spencer 1896, p. 197; Fig. 3.2). Spencer called his desert region the Eyrean and split Tate's Euronotian into a northern Torresian region and a southeastern Bassian region and ignored Tate's southwestern autochthonous region. These regions were accepted by botanists and zoologists alike and by the 1960s had established themselves in the phytogeographical and zoogeographical literature (Burbidge 1960; Main et al. 1958; Serventy and Whittell 1951). In the 2000s these zoogeographic and phytogeographic regions were renamed biomes "that are defined

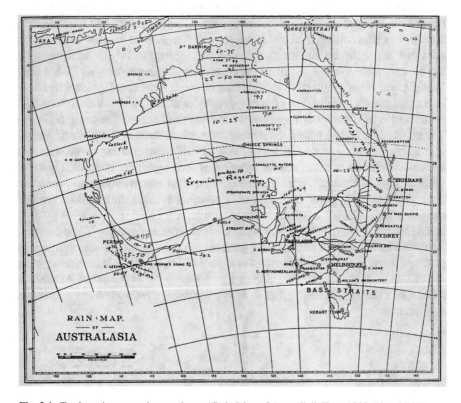

Fig. 3.1 Tate's regions superimposed on a "Rain Map of Australia" (Tate, 1889, Plate XVIII)

using a combination of climate, vegetation structure and ecophysiology" and are approximately 25 million years old when "the aseasonal-wet biome (rainforest and wet heath) gave way to the unique Australian sclerophyll biomes dominated by eucalypts, acacias and casuarinas" (Crisp et al. 2004, p. 1551; Fig. 3.3). These

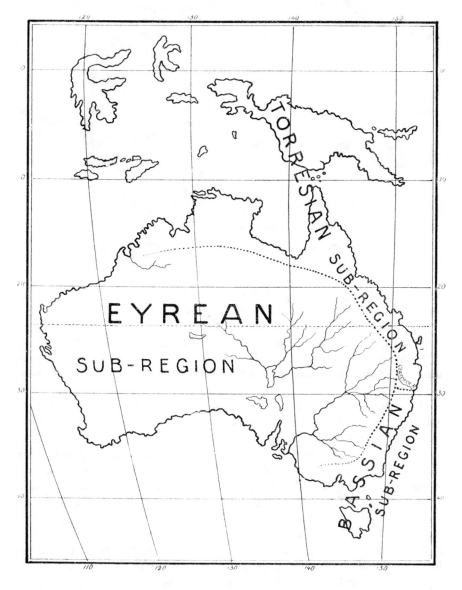

FAUNAL   SUB-REGIONS   OF   THE   AUSTRALIAN   REGION.

Fig. 3.2 Spencer's (1896) "Faunal sub-regions of the Australian region", including the Larapintine region of Tate (1896)

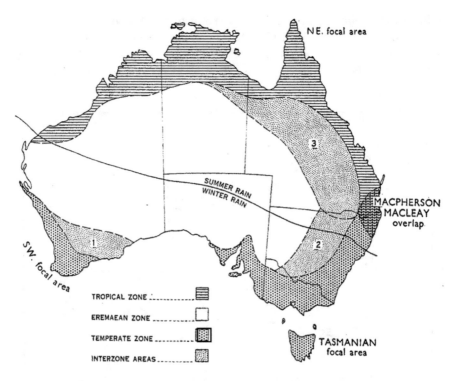

Fig. 3.3 Burbidge's zones (Burbidge, 1960, Fig. 1). Note the north-south and east-west division. (Reproduced with permission of CSIRO Publishing, Australia)

Neogene biomes, borrowed from Schodde (1989), overlapped with those of Spencer (1896) including an "aseasonal wet" biome proposed by Schodde (1989) as the older Palaeogene Tumbunan Element (akin to Tate's Oriental and Andean immigrant elements).

The main cause of Australia's biomes proposed since the 2000s was that the continent had drifted north and had become "progressively desiccated over the last 20 million years" (Byrne et al. 2008, p. 4400), suggesting that climate and a northerly tectonic shift were the main abiotic factors driving biotic diversification in the Neogene. For instance, the Australian monsoonal tropics are thought to be "a component of a single global climate system, characterized by a dominant equator-spanning Hadley cell" (Bowman et al. 2010, p. 201), while the desert regions were resultant of climate extremes during glacial and interglacial periods in the Pleistocene (Byrne et al. 2008). Moreover the "fragmentation of the eastern mesic region appears to have been driven primarily by climate change as the Australian continent drifted north" (Byrne et al. 2011, p. 1645; see Hopper and Gioia 2004 for similar explanation of the Southwestern region). While these reviews of the Australian biome do acknowledge a combination of climate and landforms as important in defining Australia's biomes, few have identified exactly what *geological* processes may have contributed towards Australia's geomorphology. Perhaps we should return to Tate (1889), or even earlier to von Ferdinand von Mueller, who noted that "to draw the species into physiographic and regional complexes must be the work of future periods, when climatic and geologic circumstances throughout Australia shall have become more extensively known" (von Mueller 1882, p. viii; also in Tate 1889, p. 312). These "geologic circumstances" have developed dramatically in the first part of the twenty-first century suggesting that "tectonic activity, as opposed to climate, exerts dominant control on bedrock erosion rate across the Australian continent [...]" and that neotectonic processes "have extensively modified the coastline of Australia, strongly influenced patterns of marine inundation, and influenced the geometry of many of Australia's streams, hill slopes, basins, and uplands over the last 5–10 Ma" (Quigley et al. 2010, p. 260). Could neotectonics, rather than solely climate, be the driver of Australia's biomes?

The myth that the abiotic causes for Australian bioregionalisation are a result of the Australian plate moving north into a wetter climatic zone needs to be deconstructed. In doing so, biologists will gain a better understanding of the tectonic processes involved in shaping the Australian landscape, in particular drainage, topography, and basin formation. Most important, we would argue, is to look beyond the notion that deformation resulting in significant change in geomorphology only occurs adjacent to convergent (collision) and divergent plate margins.

3.2 Biotectonics of Australia's Biomes

The Indo-Australian Plate (IAP) is a large plate that covers much of the Indian Ocean and Australasia, including New Zealand west of the Alpine Fault (Fig. 3.4). Since the Miocene the plate has been converging with the Eurasian and Pacific plates, creating an in situ stress field within the Australian continent. The Australian continent is a low, flat, tectonically active, yet stable, and slowly eroding continent (Quigley et al. 2010; Twidale 2011), which makes it an ideal place to study Neogene dynamic topography (Sandiford 2007). The intra-plate stress field is due to mountain building thousands of kilometres away in New Guinea, the Himalayas, and southern New Zealand. Together these active tectonic zones have led to significant tectonism in the Australian continent, particularly along its passive margins since the Miocene (Hillis et al. 2008). Moreover, crustal deformation in the Australian continent and its coastlines is "not driven by plate flexure associated with subduction" (Quigley et al. 2010, p. 246) but is the result of a combination of slab burial in the north from the Miocene and the Australian-Antarctic Discordance (AAD) region formed in the Oligocene. The southern migration of the downwelling associated with the northern slab burial and its interaction with the AAD has driven dynamic topography on the Australian continent during the Neogene (Sandiford 2007; Heine et al. 2010; Fig. 3.5). Australia's tectonism driving it northwards is far more complicated than a continent drifting into a new climatic cell. Australia's landscape has been changing due to dynamic topography as a result of tectonism rather than simple climatic changes in rainfall and temperature. The evolution of its biota also reflects these changes.

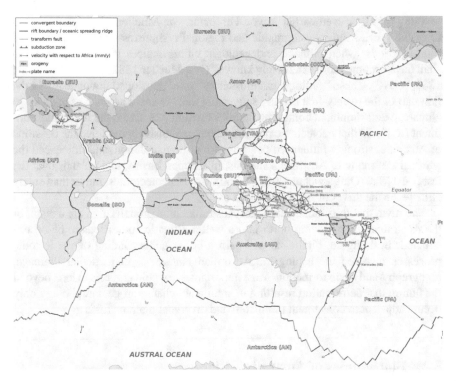

Fig. 3.4 The Indo-Australian Plate. Licensed under the Creative Commons Attribution 4.0 International license. From Wikimedia Commons https://commons.wikimedia.org/wiki/File:Tectonic_plates_boundaries_detailed_(multilanguage).svg

3.2.1 Dynamic Topography: The Neogene Tilt North

Since the Eocene the Australian continent had developed a northerly tilt, which by the early Miocene (~20 Ma) had seen a reduction in the frequent inundations of the Eucla, Murray, Otway, and Gippsland Basins (Fig. 3.6). The inundations continued to drop from ~300 m depth in the Eocene to ~70 m in the Miocene until the Nullarbor was fully emergent by ~15 Ma (Sandiford et al. 2009; Quigley et al. 2010). Eustatic sea levels would have to have been ~100–150 m above present day since the Eocene to account for these changes (Sandiford 2007; Veevers 2000), indicating that dynamic uplift was a far greater influence on these changes in inundation patterns than eustatic variation or changes in the geoid field. In addition, the northern coastline had extended further to the north, but it is difficult to identify the position of the palaeoshore line due to the presence of a late Neogene paralic lake system close by, which may have included occasional marine incursions (Doutch 1976, Sandiford 2007). The asymmetrical effect in southern and northern Australian palaeoshorelines during the mid-Miocene to late Pliocene was a result of the Australian plate tilting along a WNW-ESE tilt axis and creating a NNE-down and SSW-up vertical

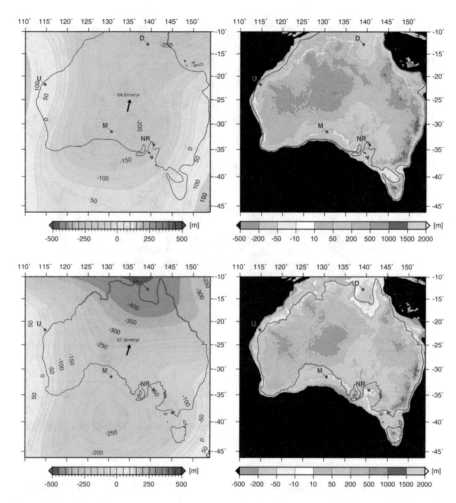

Fig. 3.5 Dynamic topography and best-fit paleogeographic reconstructions, combining dynamic topography models, eustatic sea level variations, and isostatic correction for sediments. A. ≈32 Ma (Early Oligocene, Cenozoic 3 interval); B. ≈ 15 Ma (mid-Miocene, Cenozoic 4 interval) (Heine et al. 2010, p. 143, Fig. 3.4)

movement of 10–20 m/myr (Sandiford 2007; Fig. 3.6). The tilt would had created a 300 m vertical displacement between the southwest and northern Australian continental shelves since the late Eocene (Quigley et al. 2010). The tilt and subsequent change in dynamic topography would lead one to ask "why no Nullarbor Plain along the northern Australian margin?" (Sandiford 2007, p. 158, original emphasis). The reason is that it was the tilting of the continent that resulted in a southern shelf and absence of a symmetrical northern shelf, which would have been expected if the modern geomorphology was the result of sea-level change. The northward tilt since late Miocene (~15 Ma) has left an asymmetrical palaeoshoreline, which are

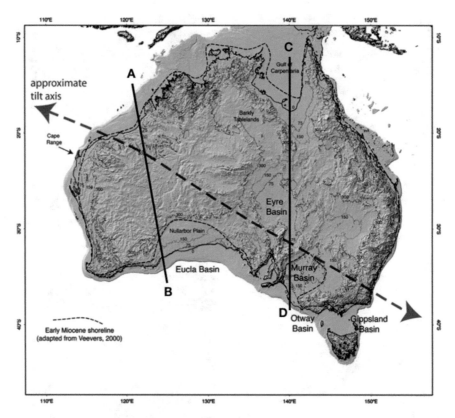

Fig. 3.6 Shaded relief image of the Australian continent and its continental shelf (Sandiford 2007, p. 155, Fig. 3.3). Note the cross-sections A–B and C–D (see Fig. 3.7)

preserved in the southern shelf (e.g. Nullarbor and Murray Basin; Hou et al. 2008) but which are generally absent along the northern shelf. This tilt had a profound effect on biogeography, giving rise to two new biomes.

3.2.2   Birth of the Australian Monsoonal Tropic and Eremaean Biomes

The Australian continent underwent significant drying during the early Miocene with the arid zone expanding from the mid-Miocene (McGowran et al. 2004). While global glacial cycles have been proposed as a cause of Australia's Neogene aridity (Byrne et al. 2011), others such as McGowran et al. (2004) also acknowledge that "it is almost equally possible to make a case for tectonism as a significant control [...] we do not yet understand the interplays between regional tectonism, tectonoeustasy,

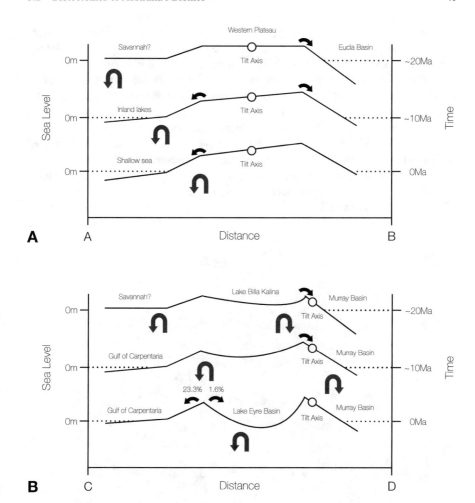

Fig. 3.7 (**a**) Cross-section from the Northwest to the Eucla Basin; (**b**) Cross-section from the Gulf of Carpentaria to the Murray Basin. (Cross-sections are shown from Fig. 3.6)

and glacioeustasy" (McGowran et al. 2004, p. 464). These tectonic processes are now becoming better understood through modelling dynamic topography processes (Heine et al. 2010). McGowran et al. (2004) concluded that the Neogene "was the time of significant collisions in the north, with far-reaching effects on palaeoceanography, climate and biogeography" (McGowran et al. 2004, p. 490). The "significant collisions in the north" may refer to the tilt, which Sandiford (2007) correctly states is rarely referenced (e.g. it is missing in McGowran et al. 2004).

The tilt north at 15 Ma does correlate with the increasing inland aridity in Australia, suggesting that it may be causal. The tilt north would effectively change the topography, from an Oligocene plateau extending further north, northwest, and northeast to the present-day coastline (Fig. 3.7). A larger plateau would mean greater drainage inland, which is presently only 0.4% of total runoff on the Western Plateau (Vardon et al. 2007, p. 651, Fig. 1; NWP and SWP in Fig. 3.8). Aridity would increase with less drainage into the reduced Western Plateau in the Neogene and Pliocene. In addition, with the tilt increasing in the Neogene and into to Pliocene, far more runoff would have occurred to the north along rivers flowing to the northwest and northeast, effectively creating a wetter environment in the northwestern and northern basins. The Neogene to Pliocene tilt alone could have created much drier centre and wetter coastal areas (without paralic lake systems or inland seas) giving rise to both the Eremaean and Australian monsoonal tropic biomes.

A southerly migration of mantle downwelling is responsible for a topographic inversion of 135 m during the late Miocene (Sandiford et al. 2009), to form the Neogene Lake Eyre and Torrens Basins, which are presently 12 m below sea level.

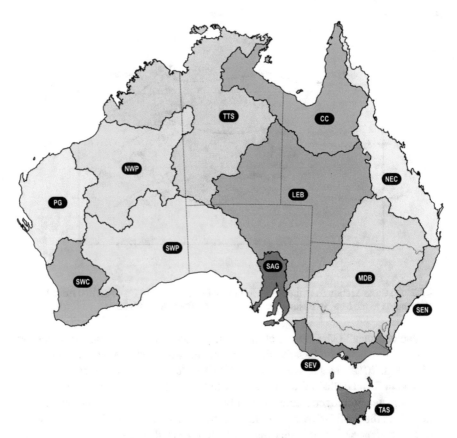

Fig. 3.8 Drainage divisions of Australia, *NEC* Northeast Coast, *SEN* Southeast Coast (NSW), *SEV* Southeast Coast (Victoria), *TAS* Tasmania, *MDB* Murray-Darling Basin, *SAG* South Australian Gulf, *SWP* Southwestern Plateau, *SWC* Southwest Coast, *PG* Pilbara-Gascoyne, *NWP* Northwestern Plateau, *TTS* Tanami-Timor Sea Coast, *LEB* Lake Eyre Basin, *CC* Carpentaria Coast. (Source Wikimedia Commons https://commons.wikimedia.org/wiki/File:Drainage_Divisions_of_Australia.svg)

The shift in the depocentre from the former Miocene Billa Kalina Basin to the Neogene Lake Eyre and Torrens Basins has resulted in significant drainage reversals, with the former Miocene southwest younging shorelines of the Eyre basin incised by east- and northeast-flowing streams (Quigley et al. 2010). The Eyre and Torrens Basins, part of the Eremaean biome, are lower and wetter, with the Eyre Basin receiving 1.6% of the total runoff, compared to the 0.4% from the Western Plateau which is higher and significantly drier. Given that both areas are in the same climate zone, they are significantly different suggesting that the Eremaean biome is in fact two separate arid biomes. One, the eastern dipping Western Plateau, a topographically high (<600 m; Czarnota et al. 2014) with no permanent water courses and two, Lake Eyre and Murray Darling Basins, topographically lower with semi-permanent to permanent rivers and lakes. Apart from being geographically distinct, each of the Eremaean biomes may have different phylogenetic histories (Byrne et al. 2008; Ebach and Murphy 2020). Further evidence for two arid biomes has been found in bioregionalisation (geospatial) studies (González-Orozco et al. 2014; Ebach et al. 2015) and in a comparative biogeographic analysis using eucalypts, *Acacia*, and *Banksia* (Murphy et al. 2019; Fig. 3.9a). A further study by Bein et al. (2020) using vertebrate distributions also had found evidence for two arid biomes (Fig. 3.9b).

a

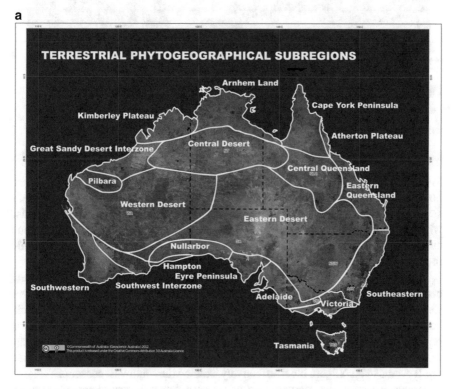

Fig. 3.9 (**a**) Map of Australia's phytogeographical subregions (Ebach et al. 2015); (**b**) Map of Australia's zoogeographic subregions (Bein et al. 2020). Note the positions of the eastern and western deserts in both studies

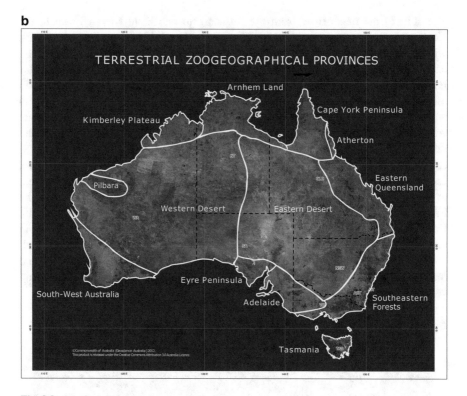

Fig. 3.9 (continued)

3.2.3 Preservation of the Temperate and Aseasonal Wet/ Tumbulan Biomes

Much of the inland temperate biome had succumbed to aridity during the Miocene; however uplift of the SW region, due to tilting, as well as the uplift of the eastern seaboard (Great Dividing Range) may be due to tectonically induced stress fields, which created significant reverse fault-bounded topographic highs within sedimentary basins in southeast Australia between 8 and 6 Ma (Quigley et al. 2010), including the Otway, Grampian, and Flinders Ranges.

The history of uplift in the Eastern Highlands has been controversial, however, because it is widely assumed that uplift began at 120 Ma and intensified during the Cenozoic (Czarnota et al. 2014). An early Cretaceous uplift, which occurred well before the onset of aridity in Australia's interior during the Neogene, would have provided the wetter environmental conditions necessary to preserve East Gondwanan aseasonal wet/Tumbulan taxa (e.g. *Nothofagus*) during later aridification. Further uplift events in the Yilgarn block effectively isolated the southwest corner of Western Australia. Rivers in the southern part of Western Australia record some 200 m of surface uplift since the mid-Eocene due to a migration of downwelling away from this part of Australia (Barnett-Moore et al. 2014). The ~200 m uplift has resulted in river reversals; the west- and south-flowing rivers changed course northwards due to the uplift of the Darling Range, and the southwest and south coasts became watersheds (Beard 1999), isolating and thereby preserving the southwest temperate biome from aridification.

3.3 New Interpretation of Neogene Biogeography

There is still much to discover about Australia's dynamic topography; however, there is enough evidence to suggest that buckling of the Australian landscape is the main driver of bedrock erosion (Quigley et al. 2010). Moreover, the motion of the Australian plate over a convecting mantle "has resulted in significant reorganization of eastern Australia catchments" (Salles and Flament 2017, p. 301). The formation of plateaux and basins and the reversal in river flows are due to neotectonics, namely, dynamic uplift due to mantle downwellings and continental tilt. These neotectonic events drove the dynamic Australian landscape and, together with climate change, created the Neogene arid and monsoonal biomes. Suggesting that neotectonics drives Australian bioregionalisation does not diminish the role of climate. Rather climate should be seen as a secondary driver in Australian bioregionalisation: landscape formation plays a significant role in shaping the distribution of scarce natural resources such as water. Without a dynamic and tilting continent, Australian bioregionalisation would look very different. Neotectonics is crucial in both shaping Neogene biomes and preserving the older Palaeogene biogeographic areas along the eastern and southwestern coasts.

The new interpretation of Australian biogeography and bioregionalisation in which neotectonics plays a vital part requires further examination. This can only be possible if biogeographers work with Earth scientists and engage with the geological literature to develop a synthesis between biology and tectonics. Australia, as Sandiford (2007) had pointed out, is an ideal place to study dynamic uplift due to the relatively flat and tectonically stable Australian continent. On other continents that are tectonically active, neotectonics may manifest itself differently; rather than creating new biomes, it may break biogeographic barriers and lead to tectonic extinction.

References

Barnett-Moore N, Flament C, Heine N, Butterworth RDM (2014) Cenozoic uplift of South Western Australia as constrained by river profile. Tectonophysics 622:186–197

Beard JS (1999) Evolution of the river systems of the south-west drainage division, Western Australia. J R Soc West Aust 82:147–164

Bein B, Ebach MC, Laffan SW, Murphy DJ, Cassis G (2020) Quantifying vertebrate zoogeographical regions of Australia using geospatial turnover in the species composition of mammals, birds. Reptiles and Terrestrial Amphibians, Zootaxa

Bowman DMJS, Brown GK, Braby MF, Brown JR, Cook LG, Crisp MD, Ford F, Haberle S, Hughes J, Isagi Y, Joseph L, McBride J, Nelson G, Ladiges PY (2010) Biogeography of the Australian monsoon tropics. J Biogeogr 37:201–216

Byrne M, Yeates DK, Joseph L, Kearney M, Bowler J, Williams MA, Cooper S, Donnellan SC, Keogh JS, Leys R, Melville J, Murphy DJ, Porch N, Wyrwoll K-H (2008) Birth of a biome: insights into the assembly and maintenance of the Australian arid zone biota. Mol Ecol 17:4398–4417

Burbidge N (1960) The phytogeography of the Australian region. Aust J Bot 8:75–211. https://doi.org/10.1071/BT9600075

Crisp MD, Cook LG, Steane DA (2004) Radiation of the Australian flora: what can comparisons of molecular phylogenies across multiple taxa tell us about the evolution of diversity in present-day communities? Philos Trans R Soc Lond Ser B Biol Sci 359:1551–1571. https://doi.org/10.1098/rstb.2004.1528

Czarnota K, Roberts GG, White NJ, Fishwick S (2014) Spatial and temporal patterns of Australian dynamic topography from River Profile Modeling. J Geophys Res Solid Earth 119. https://doi.org/10.1002/2013JB010436

Doutch HF (1976) The Karumba Basin, northeastern Australia and southern New Guinea. BMR J Aust Geol Geophys 1:131–140

Ebach MC, Murphy DL, González-Orozco CE, Miller JT (2015) A revised area taxonomy of phytogeographical regions within the Australian Bioregionalisation atlas. Phytotaxa 208:261–277

Ebach MC, Murphy DJ (2020) Carving up Australia's arid zone: a review of the bioregionalisation of the Eremaean and Eyrean biogeographic regions. Aust J Bot

González-Orozco CE, Thornhill AH, Knerr N, Laffan SW, Miller JM (2014) Biogeographical regions and phytogeography of the eucalypts. Divers Distrib 20:46–48. https://doi.org/10.1111/ddi.12129

Heine C, Müller RD, Steinberger B, DiCaprio L (2010) Integrating deep earth dynamics in paleogeographic reconstructions of Australia. Tectonophysics 483:135–150

Hillis RR, Sandiford M, Reynolds SD, Quigley MC (2008) Present-day stresses, seismicity and Neogene-to-recent tectonics of Australia's 'passive' margins: intraplate deformation controlled by plate boundary forces. In: Johnson H, Doré AG, Gatliff RW, Holdsworth R, Lundin ER, Ritchie JD (eds) The nature and origin of compression in passive margins, vol 306. Geological Society, London, Special Publications, pp 71–90

Hou B, Frakes LA, Sandiford M, Worrall L, Keeling J, Alley NF (2008) Cenozoic Eucla Basin and associated palaeovalleys, southern Australia — climatic and tectonic influences on landscape evolution, sedimentation and heavy mineral accumulation. Sediment Geol 203:112–130

Hopper SD, Gioia P (2004) The southwest Australian floristic region: evolution and conservation of a global hot spot of biodiversity. Annu Rev Ecol Evol Syst 10:399–422

McGowran B, Holdgate GR, Li Q, Gallagher SJ (2004) Cenozoic stratigraphic succession in southeastern Australia. Aust J Earth Sci 51:459–496

Main AR, Lee AK, Littlejohn MJ (1958) Evolution in three genera of Australian frogs. Evolution 12:224–233. https://doi.org/10.2307/2406033

Mueller F, von (1882) Systematic census of Australian plants. McCarron, Bird & Co., Melbourne

Murphy DJ, Ebach MC, Miller JT, Laffan SW, Cassis G, Ung V et al (2019) Do phytogeographic patterns reveal biomes or biotic regions? Cladistics. https://doi.org/10.1111/cla.12381

Quigley MC, Clark D, Sandiford M (2010) Tectonic geomorphology of Australia. Geol Soc Lond Spec Publ 346:243–265

Salles T, Flament N, Mu€ller D (2017) Influence of mantle flow on the drainage of eastern Australia since the Jurassic period. Geochem Geophys Geosyst 18:280–305. https://doi.org/10.1002/2016GC006617

Sandiford M (2007) The tilting continent: a new constraint on the dynamic topographic field from Australia. Earth Planet Sci Lett 261:152–163

Sandiford M, Quigley M, de Broekert P, Jakica S (2009) Tectonic framework for the Cainozoic cratonic basins of Australia. Aust J Earth Sci 56:5–18

Schodde R (1989) Origins, radiations, and sifting in the Australasian biota – changing concepts from new data and old. Aust Syst Bot Newsletter 60:2–11

Serventy DL, Whittell HM (1951) A handbook of the birds of Western Australia. University of Western Australia Press, Perth

Spencer WB (1896) Report on the work of the horn scientific expedition to Central Australia: part I. introduction, narrative, summary of results, supplement to zoological report, map'. Melville, Mullen & Slade, Melbourne

Tate R (1889) On the influence of physiological changes in the distribution of life in Australia. In: Liversidge A, Etheridge R (eds) Report of the first meeting of the Australasian Association for the Advancement of science. August and September 1888, Sydney. Australasian Association for the Advancement of Science, Sydney, pp 312–326

Tate R (1896) Section II. – Botany. In: Spencer WB (Ed.) Report on the work of the horn scientific expedition to Central Australia: part I. introduction, narrative, summary of results, supplement to zoological report, map', pp. 171–194. Melville, Mullen & Slade, Melbourne

Twidale CR (2011) Is Australia a tectonically stable continent? Analysis of a myth and suggested morphological evidence of tectonism. Prog Phys Geogr 35(4):493–515

Vardon M, Lenzen M, Peevor S, Creaser M (2007) Water accounting in Australia. Ecol Econ 61(4):650–659

Veevers JJ (2000) Billion-year earth history of Australia and neighbours in Gondwanaland. GEMOC Press, Sydney

Chapter 4
Biotectonics: Making and Breaking Barriers

Abstract We investigate some exemplar regions where the integration of neotectonics and dynamic topography had created unique biogeographic areas, namely, transition zones. We also term two new concepts, Marginal Plate Biotectonics and Intra-plate Biotectonics, to distinguish between different types of tectonic and biogeographic interactions. A preliminary analysis reveals that transitions zones along tectonic margins (subduction, transverse) share greater similarity to each other than they do with those found within plates. We also propose a case for biotectonic extinction and discuss the future of biotectonics.

4.1 Dynamic Topography as a Biogeographic Barrier Breaker

Unlike biomes, biogeographic areas form over much longer times, and no current biogeographic region would be a result of neotectonics (i.e. formed in the Neogene). Australian biogeographic areas, for instance, are old, while biomes are relatively young (Ebach 2017). For example, many Australian taxa, such as the eucalypts, *Acacia*, and *Banksia*, have fossil records extending into the Palaeogene (Murphy et al. 2019) while other taxa such as *Agathis* and other Araucariaceae extending back further into the Mesozoic. Breaking or degrading biogeographic barriers, that is, creating areas of mixing (i.e. overlap or transition), can be done within much shorter time frames (Ferro and Morrone 2014). The mixing of biota along biogeographic boundaries due to intra-plate neotectonism or the collision of two or more plates or terranes are examples of biotectonics.

4.1.1 Intra-plate Biotectonics

We propose *Intra-plate Biotectonics* (IPB) as a general term for biotic barrier formation or degradation due to dynamic topography away from active tectonic margins (i.e. passive margins or intra-plate processes). Biogeographic areas separated by barriers such as plateaux, inland seas, or marine transgressions may be con-

M. C. Ebach, B. Michaux, *Biotectonics*, SpringerBriefs in Evolutionary
Biology, https://doi.org/10.1007/978-3-030-51773-1_4

nected due to dynamic uplift creating marine regressions in cases of sea barriers or dynamic depression in areas where elevation is a barrier. Two biotic transition zones, the Chinese and Afro-Arabian, and the drainage patterns of the Amazon basin are discussed below as examples of the effects of dynamic topography on biotic distributions.

4.1.1.1 Chinese Transition Zone (CTZ)

The Chinese Transition Zone (sensu Morrone 2015; Hermongenes de Mendonça and Ebach 2020; Fig. 4.1) marks the boundary between the Palearctic and Oriental regions. The Oriental region of southern China is wet and warm and dominated by broadleaf and bamboo forests, while the Palearctic region is drier and colder and dominated by grasslands. The very obvious bioregionalisation of China has long been recognised and has been described for many diverse groups (Norton et al. 2010; He et al. 2017; Hu et al. 2017; Huang et al. 2020), but the position of the boundary is uncertain and its location dependent on factors such as the taxa used to define it (Huang et al. 2020), and it changes through time as biotas track their preferred habitats (Norton et al. 2010). While there are significant geographical barriers between the Oriental and Palaearctic regions in the west (Himalayas) and central China (Quinling Mountains), there is no geographical barrier along the present eastern seaboard (Norton et al. 2010). The eastern seaboard occupied an area of signifi-

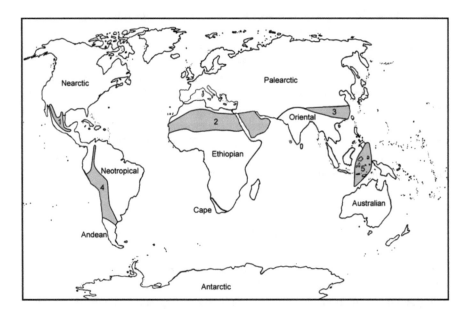

Fig. 4.1 "World biogeographical regionalisation, with indication of the regions (white) and transition zones (grey). Transition zones: (1) Mexican; (2) Saharo-Arabian; (3) Chinese; (4) South American; 5, Indo-Malayan [Wallacea]" (Morrone 2015, p. 84, Fig. 1)

cant intra-plate deformation due to downwelling that reshaped the topography of eastern China, including significant subsidence of the East China Sea Shelf Basin since the Jurassic (Cao et al. 2018). Dynamic topographic depression in the order of 1800 m along the eastern coast of China was reversed by about 70 Ma as the underlying subducting slab moved westwards relative to the overriding plate (Cao et al. 2018, Fig. 5). This uplift may have exposed areas of coastal plain allowing a zone of biotic mixing as new habitat for terrestrial organism was created. The role of local extinction of species prior to any interchange taking place has also been suggested as a mechanism that makes biotas more susceptible to invasion (Vermeij 1991), a process expected at the margins as habitats change and populations come under physiological stress (Cabezas et al. 2007; Martínez-Mota et al. 2007; Tewksbury et al. 2008). Climatic fluctuations, particularly during the Plio-Pleistocene, may have maintained and enhanced this transition zone. Further investigation into the uplift and later subsidence and other intra-plate topographic processes in Eastern China is needed to fully understand the history of the Oriental and Palearctic biogeographic barrier. The origin of the biota of southern China, which forms part of the Oriental region with India and Southeast Asia, most likely lies further back in time to the Mesozoic because north and south China are composed of separate terranes, derived from different parts of Gondwana, and only amalgamated sometime in the Triassic (Metcalfe 2013). The similarity of the southern Chinese biota to Southeast Asia reflects the earlier amalgamation of the South China block with the various terranes of Southeast Asia resulting in a shared evolutionary history and to climatic similarities that explain its persistence. The core biota of the South China craton reflects its geological and biological histories, which are different from that of the rest of China and the Palaearctic.

4.1.1.2 Saharo-Arabian Transition Zone (SATZ)

The Saharo-Arabian Transition Zone (SATZ) is an area of overlap between Palearctic and African biotas that stretches across the Sahara-Sahel, Arabian Peninsula, and the Middle East to Pakistan. This region is predominantly arid but contains a surprisingly rich endemic fauna, often of restricted geographical range, which is composed of a mixture of Palearctic and African taxa (Brito et al. 2014). According to Kappelman et al. (2003), Afro-Arabian mammal communities changed markedly at 24 Ma (Oligocene/Miocene boundary) with an influx of Eurasian migrants. Based on their study of fossil paenungulates in Ethiopia, they reported that the migrants replaced some endemics leading to their extinction (e.g. arsinotheres), while other endemics coexisted with the new migrants into the Miocene (e.g. hyracoids) or even diversified and expanded their ranges northwards into Eurasia (e.g. proboscideans). In the southern Levant and Arabian Peninsula, a "kaleidoscopic admixture of Palearctic, Paleotropic, and Saharo-Arabian-Sindian elements changes constantly during the Neogene and the Quaternary periods" (Tchernov 1992). While the great interchange occurred in the early Miocene (20–16 Ma), there had been some earlier exchanges between the Palearctic and African faunas. Similar Neogene mixing of

Palearctic and African biotas was also recorded further east in Pakistan (Flynn et al. 2016). Dates derived from molecular phylogenies provide support for both an early Miocene origin of SATZ (e.g. Amer and Kumazawa 2005; Pook et al. 2009) and earlier interchanges. For example, Wüster et al. (2008) timed the split between the Asian pit vipers (Crotalinae) and the African Viperinae between 58 and 38 Ma.

At the end of the Mesozoic, the African plate's transform motion with respect to Europe changed to NE-directed thrusting (Kley and Voigt 2008), resulting in ophio-lite obduction along the northern margin as Africa overrode intra-oceanic subduction zones in the neo-Tethys (Robertson et al. 2012) and intra-plate thrusting in central Europe, southern France, Spain, and North Africa. A widespread marine transgres-sion is also recorded from North Africa, the Arabian Peninsula, and Pakistan at this time (Ronov 1994, Bata et al. 2016). Barnett-Moore et al. (2017) argued that dynamic topography and global sea fluctuations could explain the marine inunda-tion of northern Africa and Arabia during the Late Cretaceous and the subsequent regression at the end of the Oligocene (Beauchamp et al. 1999). Barnett-Moore et al. (2017) suggested that dynamic subsidence resulted from Africa's movement away from a zone of low shear velocity (African Superplume of Moucha and Forte [2011]) and its subsequent interaction with relict subduction slabs in the neo-Tethys. As the African Plate started to interact with the European foreland, subduction rates slowed (O'Conner et al. 1999), and the effect of slab pull declined resulting in upward flexure. While the role that this subsidence played in maintaining the sepa-ration of Palearctic and Afrotropical faunas is difficult to evaluate in the light of the complex tectonic events associated with collision zones, the marine regression may have left an area open to biotic expansion from both the Palearctic and Afrotropical regions leading to a mixed biota occupying the Sahara-Sahel region.

4.1.1.3   Amazon Basin

Another example of barrier degrading due to IPB is seen in the Amazon basin (Fig. 4.2). A number of studies have suggested that large-scale subsidence and uplift during the Cenozoic of South America are not explicable in terms of tectonic defor-mation (Braun 2010; Shephard et al. 2010; Eakin and Lithgow-Bertelloni 2014; Flament et al. 2015; Dávila et al. 2019). Dávila et al. (2019) described the pericra-tonic forelands – a zone between the Andes and eastern shield regions unaffected by tectonic activity since the opening of the southwest Atlantic and the beginning of the Andean orogeny – as regions of limited relief that have been at or near sea level since at least the Cretaceous. Shephard et al. (2010) reported dynamic subsidence of <40 m/my in this region, which influenced drainage patterns causing the formation

---

Fig. 4.2 (continued) of our palaeo-topography, we do not remove any sediments or orogenic build-ing, and therefore our predictions of inundation are underestimated. The approximate extent of the Miocene Amazonian mega-wetland is illustrated by the black outline. (Adapted from ref. 29). The Purus arch and Monte Alegre arch included for reference. The red arrows delineate the sediment influx direction in the Amazonian sedimentary basins" (Shephard et al. 2012, p. 871, Fig. 1)

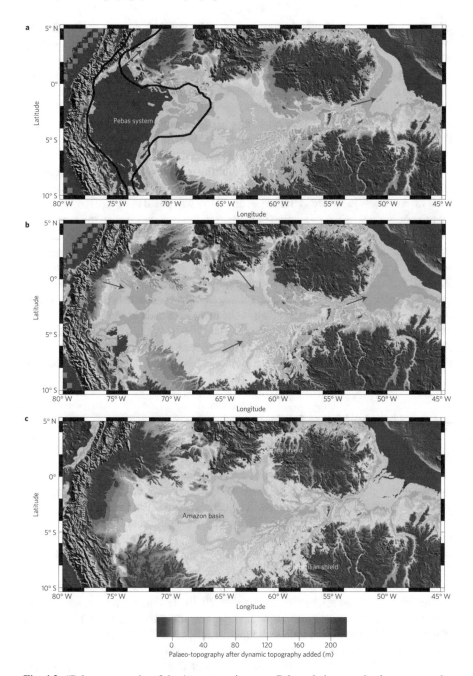

Fig. 4.2 "Palaeo-geography of the Amazon region. a–c, Palaeo-drainage and palaeo-topography of the Amazon region at 14 Myr BP (**a**), 6 Myr BP (**b**) and present-day (**c**), in a fixed South America reference frame. Long-wavelength dynamic topography signal causes significant subsidence in the central and eastern Amazon region, and uplift adjacent to the Andes. In the calculation

of an extensive inland wetland system ($\sim 10^6$ km^2) as river systems flowed westwards. As Wesselingh and Salo (2006) noted "The Pebas system had a variety of influences over the evolution of Miocene and modern Amazonian biota; it formed a barrier for the exchange of terrestrial biota, a pathway for the transition of marine biota into freshwater Amazonian environments, and formed the stage of remarkable radiations of endemic molluscs and ostracods". Antonelli et al. (2009) also discussed the role this wetland system had as a barrier to any floral interchange between the Andes and Amazonia until its disappearance in the Miocene. During the Miocene, northern South America tilted to the east as western Amazonia rebounded and dynamic subsidence migrated eastwards resulting in drainage reversal and the establishment of an eastwards flowing Amazon allowing the expansion of várzea and terra firma forests eastwards (Bicudo et al. 2019). Shephard et al. (2010) suggested that the migration of the dynamic subsidence to the east was due to the westwards movement of northern South America over subducted slabs. Others suggest that flat slab subduction brought about by the subduction of a buoyant Nazca Ridge was responsible for these changes (Eakin and Lithgow-Bertelloni 2014; Flament et al. 2015; Dávila et al. 2019). From a topographic perspective, pre-Miocene Amazon drainage had originated from the cratonic region near the northeastern coast, where it drained in a westerly direction towards the mega-wetlands of the Pebas system or Pacific (Antonelli et al. 2009; Shephard et al. 2010). With uplift adjacent to the Andes and subsidence in the east at 14 Ma, the Amazon reversed its drainage direction, flowing towards the Atlantic, thereby degrading the northeastern biotic barrier.

4.1.2   Marginal Plate Biotectonics

Marginal Plate Biotectonics (MPB) is used herein as a general term for biotic barrier formation or degradation on active tectonic margins (i.e. plate convergent and divergent margin). There are three areas of biotic overlap or transition formed at modern convergent margins, which will be discussed to illustrate MPB.

Wallacea, a biotic overlap zone bounded by the Wallace and Weber Lines and located along the "Ring of Fire" in Southeast Asia, is well-known for its tectonic origin. This overlap zone is a complex of Australian and Sundaic microcontinental fragments, island arcs, and marginal basins brought together over the past 45 Ma and compressed as a northwards-migrating Australia collided with Sundaland and the Philippine Sea/Pacific plate margin (Michaux 2019). Terrane formation, translation, and sometime amalgamation brought together distantly related Oriental and Australasian biota in places such as Sulawesi and Timor. The Wallace Line dates to the first initial barrier at ~45 Ma when southwest Sulawesi separated from eastern Borneo with the opening of the Makassar Strait (Michaux, 1996). As Australia moved northwards, continental blocks detached from both the Australian plate and Sundaland block mixed within the Wallacean collision zone (Metcalfe 2013). As a result of the movement of Sundaic/Asian terranes and their collision with the leading edge of Australia during the Neogene, the southern boundary of Wallacea was moved closer towards Australia (Weber's Line) (King and Ebach 2018; Michaux 2019).

A similar collision event between the North American, Cocos, Caribbean, and South American plates resulted in a transition or overlap of North and South American and Caribbean biota. This region of high biodiversity stretches from parts of southwest USA into Mexico and southwards through Central America to the Nicaraguan lowlands. The zone is structured into northern and southern sections, with the northern area showing lower degrees of endemism and stronger North American phylogenetic connections and a southern area with higher levels of endemism and relationships with South American and African taxa (Halffter 1987; Marshall and Liebherr 2000; Morrone 2006). The Mexican Transition Zone (MTZ) represents the northern part of this overlap, with allochthonous terranes being scattered along the northwestern margin of South America and the southern margin of North America, such as the Chocó terrane, located in northwest South America, which was possibly part of Cuba (Echeverry et al. 2012). The MTZ consists of three allochthonous terranes, namely, Southern Chortís and Siuna terranes, as well as the Santa Elena, Esperanza, and Matapalo complex, the first two having a close relationship with the Gorgona terrane in northwestern South America (Echeverry et al. 2012). These allochthonous terranes in Wallacea and the Mexican Transition Zone may have moved geographically distant and unrelated biota closer together forming a mixed or transitional biota (Michaux 2010).

The Andes also contains a mixed biota consisting of cool, temperate Andean (Patagonian) and Neotropical taxa that Morrone (2006) called the South American Transition Zone (SATZ). While not as pronounced in character as Wallacea or the MTZ, the SATZ may be a result of both tectonic uplift and downwelling due to flat slab subduction (Giambiagi and Ramos 2002; Dávila et al. 2019). The connection between the SATZ and dynamic topography is a little tenuous, mostly in part due to the poor definition of the barrier between the Andean and Neotropical regions. For instance, Morrone (2015) considered the ATZ to include "Andean highlands between western Venezuela and northern Chile and central western Argentina [...] It corresponds to the boundary between the Neotropical and Andean regions, which was analysed by Rapoport (1971), who discussed the alternative placements given by different authors to the 'subtropical line' that separates both regions" (Morrone 2015: 87–88). Roig-Juñent et al. (2018), however, consider the Patagonian Steppe to represent the southern-most part of the SATZ. Given the SATZ covers such a large area over a varied geological terrain, it may seem unlikely that a single tectonic event may contribute to the dynamic topography of the SATZ. Perhaps flat slab subduction as well as intra-plate stresses, such as basin subsidence (Flament et al. 2015), could be a factor.

4.1.3 Quantifying MPB and IPB

Comparative biogeography identifies and classifies natural biogeographic areas within an area-taxonomy; however, it does not identify areas of mixing (overlap or transition zones). Transition zones are non-monophyletic (artificial) areas that contain the biota of two unrelated areas and should be regarded as artificial in the same way non-monophyletic taxa are in systematics (Ebach and Michaux 2017). These areas do however have a biotectonic signature, one that focuses on the interplay between neotectonics/tectonics and biota. This biotectonic signature would be identifiable through properties such as dynamic uplift and biotic mixing and would fall into an artificial key or classification. For instance, a biotectonic key would classify transition zones or other areas of biotic mixing (e.g. biomes) as part of MPB and IPB. To achieve a biotectonic key, the tectonostratigraphic properties of the transition zones would need to be compared for similarities, in the same way terranes are compared for geological similarities (i.e. Echeverry et al. 2012, Michaux et al. 2018). A simple binary (e.g. absence/presence) matrix in which the tectonostratigraphic characteristics of transition zones are listed and run through a cladistic analysis will reveal the relationships between these transition zones and identify them as MPB or IPB.

4.1.4 Biotectonic Analysis: Identifying MPB and IPB

In an attempt at a biotectonic analysis, eight characters were chosen to characterise a MPB and IPB biota (Table 4.1). The tectonostratigraphic characters are as follows:

1. Forearc/arc setting: indicator that the biota is at an active plate margin
2. Flat slab: indicator of rapid mantle downflow followed by uplift
3. Tectonostratigraphic terrane: indicator that the biota is at an active plate margin involving transverse faulting
4. Intra-plate deformation: indicator that the biota is away from an active plate margin and subjected to dynamic topography
5. Decreasing dynamic topography: indicator of subsidence or back titling
6. Increasing dynamic topography: indictor of uplift or forward titling
7. Marine transgression and/or regression: indicator of subsidence/uplift and sea-level fluctuations

The cladistic analysis using Lisbeth (Zaragüeta-Bagils et al. 2012; Branch and Bound search [-obb] using default options) found an intersection geogram of 2 compatible trees (i.e. a branching diagram depicting similarities between terranes). The geogram (Fig. 4.3) groups the transitions zones into MPB (MTZ, Wallacea, and STZ) and IPB (SATZ and CTZ) biota as predicted (also see Hermongenes de Mendonça and Ebach 2020). Serving as an example of a biotectonic analysis, further tectonostratigraphic characters may be used to separate out the ATZ from the MPB group. Moreover, other suspected mixed biota may also be used in this analysis to determine whether they belong to MPB and IPB areas. If so, these areas would require further investigation as to whether their biota are mixed and to what degree, bearing in mind that most boundaries between biogeographic regions, subregions, and provinces do have a minor degree of biotic mixing (Ferro and Morrone 2014).

Table 4.1 Parenthetic data matrix with 8 characters (see text) and the 5 transitions zones (TZ) identified by Morrone (2015)

Character	Character tree
Orogeny	(B,C(A,D,E))
Flat slab	(A,B,C (D,E))
Forearc/arc	(B,C(A,D,E))
Terrane	(B,C,E(A,D))
Intra-plate deformation	(A,D,E(B,C))
Decreasing DT	(C,D,E(A,B))
Increasing DT	(A,B,D(C,E))
Marine trans/regression	(B,D,E(A,C))

A Wallacea, *B* Chinese TZ, *C* Afro Arabian TZ, *D* Mexican TZ, *E* South American TZ

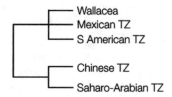

Fig. 4.3 The analysis found a single geogram with the transition zones grouped into MPB (MTZ, Wallacea and STZ) and IPB (SATZ and CTZ) as predicted

4.2   A Case for Biotectonic Extinction?

The idea that tectonics may play a part in mass extinction is not new. Biotic extinctions (e.g. biotic crises, replacement, turnover) are mostly discussed in the palaeobiogeographic literature (see Rea et al. 1990; Racki 1988; Keller 2005). For example, plate tectonics has long been considered by palaeontologists as a likely driver of extinction and replacement (e.g. Valentine and Moores 1970). A recent study, however, has provided "compelling evidence that the shifting distribution and connectivity of continental landmass, either as the primary forcer or as a complement to other macroevolutionary processes, has been a fundamental driver of long-term global Phanerozoic biodiversity patterns" (Zaffos et al. 2017, p. 5656; Fig. 4.4). Zaffos et al. (2017) is by far the most comprehensive study to empirically link continental fragmentation and collision with biodiversity. The result goes a long way to show that Earth processes, particularly plate tectonics and continental amalgamation, have a far greater impact on extinction both at the global and local levels than random events such as asteroid impacts.

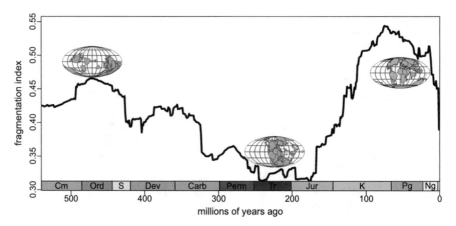

Fig. 4.4 "An index of continental block fragmentation derived from the EarthByte paleogeographic reconstruction models calculated in million-year increments. An index value of unity indicates no plates are touching; an index value of zero corresponds to the state that would be achieved if all of continental blocks were contiguous and arranged in a single mass with a minimal ratio of perimeter to area. Example Earth Byte-derived configurations (23) for the Early/Middle Ordovician (470 ma), the Middle/Late Triassic (237 ma), and Paleocene/Eocene (56 ma) are shown for visual reference. Smaller plates that do not persist throughout the entire Phanerozoic are not used in the index calculations and are grayed out in the inset maps. *Carb* Carboniferous, *Cm* Cambrian, *Dev* Devonian, *Jur* Jurassic, K, Cretaceous; *Ng* Neogene, *Ord* Ordovician, *Perm* Permian, *Pg* Paleogene, *S* Silurian, *Tr* Triassic" (Zaffos et al. 2017, Fig. 1)

At the local level, biotectonics results in area disruptions, such as transition zones (e.g. Neogene extinction of some African mammals), and in cases of localised biotic area extinctions due to dynamic topography (see Chap. 3). For instance, the highly endemic Devonian Malvinokaffric realm became extinct during the Givential (~370 Ma) with a genus-level turnover rate of 86% (Dowding and Ebach 2019). The Devonian sedimentation of the intracratonic Paraná Basin suggests a regressive littoral environment with a steep decline in shelly fossils (see Uriz et al. 2016), while globally most areas were undergoing major transgressive event (i.e. global sea-level rise; Haq and Schutter 2008). While the causes for the Malvinoklaffric extinction are clearly abiotic (i.e. draining followed by a marine transgression and new biota), the exact geological mechanisms are not fully known. If the mechanisms for extinction/replacement are biotectonic and biotectonic extinction is far more common than previously thought (e.g. Zaffos et al. 2017), then more research is required into areas not normally considered by biogeographers, such as geophysics and solid Earth dynamics. Moreover, due to better preservation of landforms and fossil mantle convection systems, the Cenozoic is the best place to find examples of Intra- and Marginal Plate Biotectonics.

4.3 Epilogue

The Australides present us with 600 Ma of biotectonic history. This history involved repeated cycles of terrane formation, translation, and amalgamation that provided a dynamic landscape upon which evolution unfolded. Vicariant events, such as basin formation, changing island arc geography (Renema et al. 2008), or mountain building (Rahbek et al. 2019), would have increased biodiversity by promoting allopatric (e.g. Turelli et al. 2001) or peripatric speciation (e.g. Williamson 1981). Range expansion or habitat modification, on the other hand, would have led to local extinction and a decrease in biodiversity (Vermeij 1991). Range expansion can occur when tectonic dispersal brings separate biotas together during terrane amalgamation or when geographic barriers are removed (e.g. Pickering et al. 1988). We conclude, therefore, that evolutionary rates should be higher in tectonically active margins than in stable cratons. High rates of evolutionary change within the biotas of orogenic belts, isolated to varying degrees from their cratonic forelands, will result in the development of distinct regionalisation. We have presented evidence to support this view.

While we expect taxa distributed in cratonic areas to show slower rates of evolutionary change, it would be wrong to think of cratonic areas as uninteresting from an evolutionary or biogeographic viewpoint. Recent studies into long amplitude topographic anomalies, manifest in flat, highly eroded continental cores far from any active margin, have shown that relatively modest elevation differences (< ±1000 m) can have a profound effect, particularly on drainage patterns. Alteration of drainage patterns can promote speciation in freshwater taxa if river systems become fragmented or cause local extinctions if barriers between river systems disappear or basins dry out. The reversal of flow direction in the Amazon and the extremely high diversity of freshwater Neotropical fish (Fernandes et al. 2004), with endemics often restricted to individual tributaries, show the potential scale of these effects. In Australia, a Neogene tilting of the entire continent has resulted in a quite profound change in the distribution of water resources. Because Australia is so dry, the way in which this scarce resource is partitioned results in effects quite out of proportion to the change in topography. Thus, the northern coast is wetter than it should be and supports a tropical monsoon biome and has a broad coastal plain and extensively developed intertidal environments flanking shallow seas providing habitat for a wide range of organisms. In contrast, the southern part of the continent is dry and has coast characterised by deep water and steep cliffs.

Neotectonics is still in its infancy, so too is a truly *historical* biogeography that takes changes in geography into account, not only on a coarse scale seen at tectonically active margins but also the subtler effects of processes such as dynamic topography. Our attempts to synthesise current neotectonic hypotheses with biogeography as the interdisciplinary field of biotectonics are hampered by a lack of data such as cladograms and fossil distributions and a lack of communication between Earth scientists and biogeographers at the operational level. There have been attempts to rectify this disparity by providing online tools that integrate palaeogeographic and paleobiological databases and other community-driven initiatives (Wright et al. 2013). We suggest that using oversimplified tectonic scenarios and poorly resolved or phylogenetically restricted taxon cladograms will not move the synthesis forwards.

References

Amer SAM, Kumazawa Y (2005) Mitochondrial DNA sequences of the afro-Arabian spiny-tailed lizards (genus *Uromastyx*; family Agamidae): phylogenetic analyses and evolution of gene arrangements. Biol J Linn Soc 85:247–260

Antonelli A, Nylander JAA, Persson C, Sanmartin I (2009) Tracing the impact of the Andean uplift on Neotropical plant evolution. Proc Natl Acad Sci U S A 106:9749–9754

Barnett-Moore N, Hassan R, Müller RD, Williams SE, Flament N (2017) Dynamic topography and eustasy controlled the paleogeographic evolution of northern Africa since the mid-cretaceous. Tectonics 36:929–944

Bata T, Parnell J, Samaila NK, Haruna AI (2016) Anomalous occurrence of cretaceous placer deposits: a review. Earth Atmos Sci 1:1–13

Beauchamp W, Allmendinger RW, Barazangi M (1999) Inversion tectonics and the evolution of the high Atlas Mountains, Morocco, based on a geological-geophysical transect. Tectonics 18:163–184

Bicudo TC, Sacek V, de Almeida RP, Bates JM, Ribas CC (2019) Andean Tectonics and mantle dynamics as a pervasive influence on Amazonian ecosystem. Sci Rep 9:16879. https://doi.org/10.1038/s41598-019-53465-y

Braun J (2010) The many surface expressions of mantle dynamics. Nat Geosci 3:825–833

Brito JC, Godinho R, Martínez-Freiría F, Pleguezuelos JM, Rebelo H, 18 others (2014) Unravelling biodiversity, evolution and threats to conservation in the Sahara-Sahel. Biol Rev 89:215–231

Cabezas S, Blas J, Marchant TA, Moreno S (2007) Physiological stress levels predict survival probabilities in wild rabbits. Hormones and Behavior 51:313–320

Cao X, Flament N, Müller D, Li S (2018) The dynamic topography of eastern China since the latest Jurassic period. Tectonics 37:1274–1291

Dávila F, Ávila P, Martina F, Canelo HN, Nóbile JC, Collo G, Nassif FS, Ezpeleta M (2019) Measuring dynamic topography in South America. In: Horton BK, Folguera A (eds) Andean tectonics. Elsevier, Amsterdam, pp 35–66

Dowding EM, Ebach MC (2019) Evaluating Devonian bioregionalization: quantifying biogeographic areas. Paleobiology 45(4):636–651. https://doi.org/10.1017/pab.2019.30

Eakin CM, Lithgow-Bertelloni DFM (2014) Influence of Peruvian flat-subduction dynamics on the evolution of western Amazonia. Earth Planet Sci Lett 404:250–260

Ebach MC (2017) Reinventing Australasian biogeography. CSIRO Publishing, Clayton

Ebach MC, Michaux B (2017) Establishing a framework for a natural area taxonomy. Acta Biotheor 65:167–177

Echeverry A, Silva-Romo G, Morrone JJ (2012) Tectonostratigraphic terrane relationships: a glimpse into the Caribbean under a cladistic approach. Palaeogeogr Palaeoclimatol Palaeoecol 353-355:87–92

Fernandes CC, Podos J, Lundberg JG (2004) Amazonia ecology: tributaries enhance the diversity of electric fishes. Science 305:1960–1962

Ferro I, Morrone JJ (2014) Biogeographical transition zones: a search for conceptual synthesis. Biol J Linn Soc 113:1–12

Flament N, Gurnis M, Müller RD, Bower DJ, Husson L (2015) Influence of subduction history on south American topography. Earth Planet Sci Lett 430:9–18

Flynn LJ, Pilbeam D, Barry JC, Morgan ME, Razaa SM (2016) Siwalik synopsis: a long stratigraphic sequence for the later Cenozoic of South Asia. Comptes Rendus Palevol 15:877–887

Giambiagi LB, Ramos VA (2002) Structural evolution of the Andes in a transitional zone between flat and normal subduction (33 30′–33 45′ S), Argentina and Chile. Journal of South American Earth Sciences 15:101–116

Halffter G (1987) Biogeography of the montane entomofauna of Mexico and Central America. Annu Rev Entomol 32:95–114

Haq BU, Schutter SR (2008) A chronology of Paleozoic sea-level changes. Science 322:64–68

He J, Kreft H, Gao E, Wang Z, Jiang H (2017) Patterns and drivers of zoogeographical regions of terrestrial vertebrates in China. J Biogeogr 44:1172–1184

Hu R, Wen C, Gu Y, Wanga H, Gu L, Shi X, Zhong J, Wei M, He F, Lu Z (2017) A bird's view of new conservation hotspots in China. Biol Conserv 211:47–55

Huang C, Ebach MC, Ahyong ST (2020) Bioregionalisation of the freshwater zoogeographical areas of mainland China. Zootaxa 4742:271–298

Kappelman J, Rasmussen DT, Sanders WJ, Feseha M, Bown T, Copeland P, Crabaugh J, Fleagle J, Glantz M, Gordon A, Jacobs B (2003) Oligocene mammals from Ethiopia and faunal exchange between Afro-Arabia and Eurasia. Nature 426:549–552

Keller G (2005) Impacts, volcanism and mass extinction: random coincidence or cause and effect? Aust J Earth Sci 52:725–757

King AR, Ebach MC (2018) A novel approach to time-slicing areas within biogeographic-area classifications: Wallacea as an example. Aust Syst Bot 30(6):495–512

Kley J, Voigt T (2008) Late cretaceous intraplate thrusting in Central Europe: effect of Africa-Iberia-Europe convergence, not alpine collision. Geology 36:839–842

Marshall CJ, Liebherr JK (2000) Cladistic biogeography of the Mexican transition zone. J Biogeogr 27:203–216

Martínez-Mota R, Valdespino C, Sánchez-Ramos MA, Serio-Silva JC (2007) Effects of forest fragmentation on the physiological stress response of black howler monkeys. Animal Conservation 10:374–379

Metcalfe I (2013) Gondwana dispersion and Asian accretion: tectonic and palaeogeographic evolution of eastern Tethys. J Asian Earth Sci 66:1–33

Michaux B (1996) The origin of Southwest Sulawesi and other Indonesian terranes: a biological view. Palaeogeography Palaeoclimate Palaeogeography 122:167–183

Michaux B (2010) Biogeology of Wallacea: geotectonic models, areas of endemism, and natural biogeographical units. Biol J Linn Soc 101:193–212

Michaux B (2019) Biogeology: evolution in a changing landscape. CRC Press, Florida

Michaux B, Ebach M, Dowding EM (2018) Cladistic methods as a tool for terrane analysis: a New Zealand example. N Z J Geol Geophys 61:127–135

Morrone JJ (2006) Biogeographic areas and transition zones of Latin America and the Caribbean Islands based on Panbiogeographic and Cladistic analyses of the Entomofauna. Annu Rev Entomol 51:467–494

Morrone JJ (2015) Biogeographical regionalisation of the world: a reappraisal. Aust Syst Bot 28:81–90

Moucha R, Forte AM (2011) Changes in African topography driven by mantle convection. Nat Geosci 4:707–712

Murphy DJ, Ebach MC, Miller JT, Laffan SW, Cassis G, Ung V, Thornhill AH, Knerr N, Tursky ML (2019) Do phytogeographic patterns reveal biomes or biotic regions? Cladistics. https://doi.org/10.1111/cla.12381

Norton CJ, Jin C, Wang Y, Zhang Y (2010) Rethinking the palearctic-oriental biogeographic boundary in quaternary China. In: Norton CJ, Braun DR (eds) Asian paleoanthropology: from Africa to China and beyond. Springer, BV, p 81–100

O'Conner JM, Stoffers P, van den Bogaard P, McWilliams M (1999) First seamount age evidence for significantly slower African plate motion since 19 to 30 ma. Earth Planet Sci Lett 171:575–589

Pickering KT, Bassett MG, Siveter DJ (1988) Late Ordovician–early Silurian destruction of the Iapetus Ocean: Newfoundland, British Isles and Scandinavia—a discussion. Trans R Soc Edinb Earth Sci 79:361–382

Pook CE, Joger U, Stümpel N, Wister W (2009) When continents collide: phylogeny, histori-cal biogeography and systematics of the medically important viper genus *Echis* (Squamata: Serpentes: Viperidae). Mol Phylogenet Evol 53:792–807

Racki G (1988) Frasnian–Famennian biotic crisis: undervalued tectonic control? Palaeogeography, Palaeoclimatology, Palaeoecology 141:177–198

Rahbek C, Borregaard MK, Antonelli A et al (2019) Building mountain biodiversity: geological and evolutionary processes. Science 365:1114–1119

Rea DK, Zachos JC, Owen RM, Gingerich P (1990) Global change at the Paleocene-Eocene bound-ary: climatic and evolutionary consequences of tectonics events. Palaeogeogr Palaeoclimatol Palaeoecol 79:117–128

Renema W, Bellwood DR, Braga JC et al (2008) Hopping hotspots: global shifts in marine biodi-versity. Science 321:654–657

Robertson AHF, Parlak O, Ustaömer T (2012) Overview of the Palaeozoic–Neogene evolution of Neotethys in the eastern Mediterranean region (southern Turkey, Cyprus, Syria). Pet Geosci 18:381–404

Roig-Juñent SA, Griotti M, Domínguez MC, 13 others (2018) The Patagonian steppe biogeo-graphic province: Andean region or south American transition zone? Zool Scr 47:623–629

Ronov AB (1994) Phanerozoic transgressions and regressions on the continents: a quantitative approach based on areas flooded by the sea and areas of marine and continental deposition. Am J SciAm J Sci 294:777–801

Shephard G, Müller R, Liu L, Gurnis M (2010) Miocene drainage reversal of the Amazon River driven by plate–mantle interaction. Nat Geosci 3:870–875

Shephard GE, Liu L, Müller RD, Gurnis M (2012) Dynamic topography and anomalously negative residual depth of the Argentine Basin. Gondwana Research 22:658–663

Tchernov E (1992) The afro-Arabian component in the Levantine mammalian fauna – a short biogeographical review. Isr J Zool 38:155–192

Tewksbury JJ, Huey RB, Deutsch CA (2008) Putting the heat on tropical animals. Science 320:1296–1297

Turelli M, Barton NH, Coyne JA (2001) Theory and speciation. Trends Ecol Evol 16:330–343

Valentine JW, Moores EM (1970) Plate-tectonic regulation of faunal diversity and sea level: a model. Nature 228:657–659

Vermeij GJ (1991) When biotas meet: understanding biotic interchange. Science 253:1099–1104

Uriz NJ, Cingolani CA, Basei MAS, Blanco G, Abre P, Portillo NS, Siccardi A (2016) Provenance and paleogeography of the Devonian Durazno group, southern Parana Basin in Uruguay. J S Am J Earth Sci 66(C):248–267. https://doi.org/10.1016/j.jsames.2016.01.002

Wüster W, Peppin L, Pook CE, Walker DE (2008) A nesting of vipers: phylogeny and historical biogeography of the Viperidae (Squamata: Serpentes). Mol Phylogenet Evol 49:445–459

Wesselingh FP, Salo JA (2006) Miocene perspective on the evolution of the Amazonian biota. Scr Geol 133:439–458

Williamson PG (1981) Palaeontological documentation of speciation in Cenozoic molluscs from Turkana Basin. Nature 293:437–443

Wright N, Zahirovic S, Müller RD, Seton M (2013) Towards community-driven paleogeographic reconstructions: integrating open-access paleogeographic and paleobiology data with plate tec-tonics. Biogeosciences 10:1529–1541

Zaffos A, Finnegan S, Peters SE (2017) Plate tectonic regulation of biodiversity. Proc Natl Acad Sci 114(22):5653–5658. https://doi.org/10.1073/pnas.170229711

Glossary

Areagram A classification of areas, based on the **cladogram** of a single taxon, usually depicted as a branching diagram or as a set of parentheses

Australides A Neoproterozoic to Late Mesozoic orogenic belt found along the palaeo-Pacific and Iapetus margins

Biogeography A multidisciplinary science that uses questions, theories, and methods from the practitioners' field in order to answer question about organismal distribution (e.g. evolutionary biologist would ask a question about the ancestral distributions of living taxa)

Biome A geographical area defined by existing climate and vegetation type (e.g. tropical rainforest), rather than the distribution of a **biota**

Biogeographic area A **biotic area** that falls within a hierarchy of areas (e.g. Southwest is a subregion of the Australian region).

Bioregionalisation (biogeographic regionalisation) The practise of dividing the world into spatial units based on their biotic and abiotic characters

Biotectonics The study of the impact of tectonics on biotic evolution and biogeoregionalisation

Biota The plants and animals living in a defined geographical area

Cladistics A theory and method of biological classification that aims to find a natural classification

Craton A region of old and tectonically stable continental crust

Down- and upwellings Mantle convection cells where colder and denser material sinks and hotter, less dense material rises

Dynamic topography Surface features formed in response to mantle convection

Geogram The relationships between abiotic geographical or geological areas based on the similarity of their parts (e.g. rock types, age)

Geoid The hypothetical surface of the Earth (equivalent to a mean sea level) that is dependent on gravity and rotation alone and takes no account of any non-uniform density distribution

M. C. Ebach, B. Michaux, *Biotectonics*, SpringerBriefs in Evolutionary Biology, https://doi.org/10.1007/978-3-030-51773-1

Intra-plate biotectonics A term for biotic barrier formation or degradation caused by dynamic topographic processes in areas away from active tectonic margins

Marginal plate biotectonics A term for biotic barrier formation or degradation on active tectonic margins

Neotectonics Neogene intra-plate deformation due to the down- and upwellings in the mantle that leads to dynamic topography and landscape development

Taxon An organism or group of organisms with a hierarchical classification (taxonomy)

Tectonostratigraphic terranes Fault-bounded geologic entities of regional extent, each characterized by a geologic history distinct from that of neighbouring terranes

Transition zone The geographic space where two biotic areas overlap. Transition zones are not considered to be natural areas. Not to be confused with the geological term for the area between the lower and upper mantle

Index

© The Author(s), under exclusive license to Springer Nature Switzerland AG 2020   65
M. C. Ebach, B. Michaux, *Biotectonics*, SpringerBriefs in Evolutionary Biology,
https://doi.org/10.1007/978-3-030-51773-1

Printed in the United States
By Bookmasters